Key Papers

CYBERNETICS

Key Papers

Cybernetics

edited by C. R. Evans, B.A., Ph.D.
and A. D. J. Robertson, B.A.

Autonomics Division,
National Physical Laboratory,
Teddington, England

UNIVERSITY PARK PRESS
Baltimore, Maryland
and
Manchester, England
1968

U.S.A.:   UNIVERSITY PARK PRESS
Chamber of Commerce Building
Baltimore, Maryland 21202

ENGLAND: UNIVERSITY PARK PRESS (U.K.)
220 Wilmslow Road
Manchester, 14

First published by
Butterworth & Co. (Publishers) Ltd.

First Printing 1968
Second Printing 1970

Suggested U.D.C. number 007
,,   *additional numbers* 621-50
621.391
Library of Congress Catalog Card Number 68-55416
Standard Book Number 8391-0015-9

Made and printed by offset litho in Great Britain by
Spottiswoode, Ballantyne & Co. Ltd., London & Colchester

INTRODUCTION

As with our previous collection in this series*, this book is aimed at two broad classes of readers – the 'Beginners' and the 'Experts'. Into the former category we include undergraduates in all scientific disciplines, but most particularly those in physiology, psychology, medicine, physics, mathematics and engineering, for whom a grasp of the fundamentals of cybernetics and its brief history should be an absolute essential. In the second category we include postgraduate students, research workers, and university staff who will be familiar with many of the papers included, but who may find this a useful and pleasurable library volume.

As with our previous anthology, (but unlike most other collections of Readings), this book could quite sensibly be read from cover to cover. Treated in this fashion it will serve as a somewhat *outré* and challenging introductory textbook. On the other hand, each of the papers included has its own particular merit, and all are self-explanatory as they stand. Students are urged however, to take up the 'Further Readings' recommended for they have been specially chosen to complement the basic material in the text.

Key Papers: Brain Physiology and Psychology (1966). Ed. by C. R. Evans and A. D. J. Robertson. Butterworths, London, and University of California Press, Los Angeles.

CONTENTS

ON THE MATHEMATICAL PRINCIPLES OF THE ANALYTICAL ENGINE

Cybernetics is generally thought of as a brand new science—or at any rate as a recent amalgam of a number of older sciences. It is true that the word, or the form of the word was coined within the past few decades, and the subject barely made any significant impact until shortly after the last war, when the sudden advance of electronic engineering made vast digital and analogue computers possible. In fact, as this introductory paper shows, the principles and logic of calculating machines were available long before the coining of the word cybernetics, and as far back as the early part of the nineteenth century. To read this account by the Italian engineer, L. F. Menabrea, of Babbage's calculating machine with its punched cards, warning bells and automatic stops and then to realize that all this was taking place before the development of steam railway travel, is a chastening and slightly unsettling experience. At the same time there is something extremely coherent and explicit about Menabrea's style of presentation which suggests unhappy comparisons with the general level of scientific writing today. The footnotes which we have included because they have a special significance are by that intriguing character, the Countess of Lovelace, who was a contemporary of Babbage and who quite clearly understood the nature of his invention and its implications. (Continued on page 288).

Further Reading
Technical details of complex modern calculating machines do not make good reading, so the further reading we recommend takes up the philosophical questions raised by the varieties of the 'Thinking Machines' argument.

Turing, A. M. (1966). 'Can a Machine Think?' In *Key Papers: Brain Physiology and Psychology*. Ed. by C. R. Evans and A. D. J. Robertson. Butterworths, London.

Gunderson, K. (1964). 'The Imitation Game'. *Mind* 73, 290.

ON THE MATHEMATICAL PRINCIPLES OF THE ANALYTICAL ENGINE

L. F. Menabrea

THIS TEXT is reprinted directly from pages 666-731 of *Taylor's Scientific Memoirs*, Vol. III.

ARTICLE XXIX

Sketch of the Analytical Engine invented by Charles Babbage, *Esq. By* L. F. MENABREA, *of Turin, Officer of the Military Engineers.*

(From the *Bibliothéque Universelle de Généve, No.* 82. October, 1842)

[Before submitting to our readers the translation of M. Membrea's memoir 'On the Mathematical Principles of the ANALYTICAL ENGINE' invented by Mr. Babbage, we shall present to them a list of the printed papers connected with the subject, and also of those relating to the Difference Engine by which it was preceded.

For information on Mr. Babbage's '*Difference* Engine,' which is but slightly alluded to by M. Menabrea, we refer the reader to the following sources:

1. Letter to Sir Humphrey Davy, Bart., P.R.S., on the Application of Machinery to Calculate and Print Mathematical Tables. By Charles Babbage, Esq., F.R.S. London, July 1822. Reprinted, with a Report of the Council of the Royal Society, by order of the House of Commons, May, 1823.

2. On the Application of Machinery to the Calculation of Astronomical and Mathematical Tables. By Charles Babbage, Esq.—Memoirs of the Astronomical Society, vol. i, part 2. London, 1822.

3. Address to the Astronomical Society by Henry Thomas Colebrooke, Esq., F.R.S., President, on presenting the first Gold Medal of the Society to Charles Babbage, Esq., for the invention of the Calculating Engine.—Memoirs of the Astronomical Society. London, 1822.

4. On the Determination of the General Term of a New Class of Infinite Series. By Charles Babbage, Esq.—Transactions of the Cambridge Philosophical Society.

5. On Mr. Babbage's New Machine for Calculating and Printing Mathematical Tables.—Letter from Francis Baily, Esq., F.R.S., to M. Schumacher. No. 46, Astronomische Nachrichten. Reprinted in the Philosophical Magazine, May, 1824.

6. On a Method of expressing by Signs the Action of Machinery. By Charles Babbage, Esq.—Philosophical Transactions. London, 1826.

7. On Errors common to many Tables of Logarithms. By Charles Babbage, Esq.—Memoirs of the Astronomical Society, London, 1827.

8. Report of the Committee appointed by the Council of the Royal Society to consider the subject referred to in a communication received by them from the Treasury respecting Mr. Babbage's Calculating Engine, and to report thereon. London, 1829.

9. Economy of Manufactures, chap. xx. 8vo. London, 1832

10. Article on Babbage's Calculating Engine.—Edinburgh Review, July, 1834. No. 120, vol. lix.

The present state of the Difference Engine, which has always been the property of Government, is as follows: The drawings are nearly finished, and the mechanical notation of the whole, recording every motion of which it is susceptible, is completed. A part of that Engine, comprises sixteen figures, arranged in three orders of differences, has been put together, and has frequently been used during the last eight years. It performs its work with absolute precision. This portion of the Difference Engine, together with all the drawings, are at present deposited in the Museum of King's College, London.

Of the ANALYTICAL ENGINE, which forms the principle object of the present memoir, we are not aware that any notice has hitherto appeared, except a Letter from the Inventor to M. Quetelet, Secretary to the Royal Academy of Sciences at Brussels, by whom it was communicated to that body. We subjoin a translation of this Letter, which was itself a translation of the original, and was not intended for publication by its author.

Royal Academy of Sciences at Brussels. General Meeting of the 7th and 8th of May, 1835

"A Letter from Mr. Babbage announces that he has for six months been engaged in making the drawings of a new calculating machine of far greater power than the first.

" 'I am myself astonished,' says Mr. Babbage, 'at the power I have been enabled to give this machine; a year ago I should not have believed this result possible. This machine is intended to contain a hundred variables (or numbers susceptible of changing); each of these numbers may consist of twenty-five figures, $v_1, v_2, \ldots v_n$ being any numbers whatever, n being less than a hundred; if $f(v_1, v_2, v_3, \ldots v_n)$ be any given function which can be formed by addition, subtraction, multiplication, division, extraction of roots, or elevation to powers, the machine will calculate its numerical value; it will afterwards substitute this value in the place of v, or of any other variable, and will calculate this second function with respect to v. It will reduce to tables almost all equations of finite differences. Let us suppose that we have observed a thousand values of a, b, c, d, and that we wish to calculate them by the formula

$$p = \sqrt{\frac{a+b}{cd}}$$

the machine must be set to calculate the formula; the first series of the values of a, b, c, d, must be adjusted to it; it will then calculate them, print them, and reduce them to zero; lastly, it will ring a bell to give notice that a new set of constants must be inserted. When there exists a relation between any number of successive coefficients of a series, provided it can be expressed as has already been said, the machine will calculate them and make their terms known in succession; it may afterwards be disposed so as to find the value of the series for all the values of the variable.'

"Mr. Babbage announces, in conclusion, 'that the greatest difficulties of the invention have already been surmounted, and that the plans will be finished in a few months.'"

In the Ninth Bridgwater Treatise, Mr. Babbage has employed several arguments deduced from the Analytical Engine, which afford some idea of its powers. See Ninth Bridgwater Treatise, 8vo, second edition. London, 1834.

Some of the numerous drawings of the Analytical Engine have been engraved on wooden blocks, and from these (by a mode contrived by Mr. Babbage) various stereotype plates have been taken. They comprise—

1. Plan of the figure wheels for one method of adding numbers.
2. Elevation of the wheels and axis of ditto.
3. Elevation of framing only of ditto.
4. Section of adding wheels and framing together.
5. Section of the adding wheels, sign wheels and framing complete.
6. Impression from the original wood block.
7. Impressions from a stereotype cast of No. 6, with the letters and signs inserted. Nos. 2, 3, 4 and 5 were stereotypes taken from this.
8. Plan of adding wheels and of long and short pinions, by means of which stepping is accomplished.

N.B. This process performs the operation of multiplying or dividing a number by any power of ten.

9. Elevation of long pinions in the position of addition.
10. Elevation of long pinions in the position for stepping.
11. Plans of mechanism for carrying the tens (by anticipation), connected with long pinions.
12. Section of the chain of wires for anticipating carriage.

4

13. Sections of the elevation of parts of the preceding carriage. All these were executed about five years ago. At a later period (August, 1840) Mr. Babbage caused one of his general plans (No. 25) of the whole Analytical Engine to be lithographed at Paris. Although these illustrations have not been published, on account of the time which would be required to describe them, and the rapid succession of improvements made subsequently, yet copies have been freely given to many of Mr. Babbage's friends, and were in August, 1838, presented at Newcastle to the British Association for the Advancement of Science, and in August, 1840, to the Institute of France through M. Arago, as well as to the Royal Academy of Turin through M. Plana—EDITOR.]

THOSE labours which belong to the various branches of the mathematical sciences, although on first consideration they seem to be the exclusive province of intellect, may, nevertheless, be divided into two distinct sections; one of which may be called the mechanical, because it is subjected to precise and invariable laws, that are capable of being expressed by means of the operations of matter; while the other, demanding the intervention of reasoning, belongs more specially to the domain of the understanding. This admitted, we may propose to execute, by means of machinery, the mechanical branch of these labours, reserving for pure intellect that which depends on the reasoning faculties. Thus the rigid exactness of those laws which regulate numerical calculations must frequently have suggested the employment of material instruments, either for executing the whole of such calculations or for abridging them; and thence have arisen several inventions having this object in view, but which have in general but partially attained it. For instance, the much-admired machine of Pascal is now simply an object of curiosity, which, whilst it displays the powerful intellect of its inventor, is yet of little utility in itself. Its powers extended no further than the execution of the four* first operations of arithmetic, and indeed

---

* This remark seems to require further comment, since it is in some degree calculated to strike the mind as being at variance with the subsequent passage (page 11), where it is explained that *an engine which can effect these four operations* can in fact effect *every species of calculation*. The apparent discrepancy is stronger too in the translation than in the original, owing to its being impossible to render precisely into the English tongue all the niceties of distinction which the French idiom happens to admit of in the phrases used for the two passages we refer to. The explanation lies in this: that in the one case the execution of these four operations is the *fundamental starting-point*, and the object proposed for attainment by the machine is the *subsequent combination of these* in every possible variety; whereas in the other case the execution of some *one* of these four operations, selected at pleasure, is the *ultimatum*, the sole and utmost result that can be proposed for attainment by the machine referred to,

were in reality confined to that of the two first, since multiplication and division were the result of a series of additions and subtractions. The chief drawback hitherto on most of such machines is, that they require the continual intervention of a human agent to regulate their movements, and thence arises a source of errors; so that, if their use has not become general for large numerical calculations, it is because they have not in fact resolved the double problem which the question presents, that of *correctness* in the results, united with *economy* of time.

Struck with similar reflections, Mr. Babbage has devoted some years to the realization of a gigantic idea. He proposed to himself nothing less than the construction of a machine capable of executing not merely arithmetical calculations, but even all those of analysis, if their laws are known. The imagination is at first astounded at the idea of such an undertaking; but the more calm reflection we bestow on it, the less impossible does success appear, and it is felt that it may depend on the discovery of some principle so general, that if applied to machinery, the latter may be capable of mechanically translating the operations which may be indicated to it by algebraical notation. The illustrious inventor having been kind enough to communicate to me some of his views on this subject during a visit he made at Turin, I have, with his approbation, thrown together the impressions they have left on my mind. But the reader must not expect to find a description of Mr. Babbage's engine; the comprehension of this would entail studies of much length; and I shall endeavour merely to give an insight into the end proposed, and to develop the principles on which its attainment depends.

I must first premise that this engine is entirely different from that of which there is a notice in the "Treatise on the Economy of Machinery," by the same author. But as the latter gave rise* to the idea of the engine in question, I consider it will be a useful preliminary briefly to recall what were Mr. Babbage's first essays, and also the circumstances in which they originated.

and which result it cannot any further combine or work upon. The one *begins* where the other *ends*. Should this distinction not now appear perfectly clear, it will become so on perusing the rest of the Memoir, and the Notes that are appended to it.—Note by Translator.

* The idea that the one engine is the offspring and has grown out of the other, is an exceedingly natural and plausible supposition, until reflection reminds us that no *necessary* sequence and connexion need exist between two such inventions, and that they *may* be wholly independent. M. Menabrea has shared this idea in common with persons who have not his profound and accurate insight into the nature of either engine.

It is well known that the French government, wishing to promote the extension of the decimal system, had ordered the construction of logarithmical and trigonometrical tables of enormous extent. M. de Prony, who had been entrusted with the direction of this undertaking, divided it into three sections, to each of which were appointed a special class of persons. In the first section the formulae were so combined as to render them subservient to the purposes of numerical calculation; in the second, these same formulae were calculated for values of the variable, selected at certain successive distances; and under the third section, comprising about eighty individuals, who were most of them acquainted with the first two rules of arithmetic, the values which were intermediate to those calculated by the second section were interpolated by means of simple additions and subtractions.

An undertaking similar to that just mentioned having been entered upon in England, Mr. Babbage conceived that the operations performed under the third section might be executed by a machine; and this idea he realized by means of mechanism, which has been in part put together, and to which the name Difference Engine is applicable, on account of the principle upon which its construction is founded. To give some notion of this, it will suffice to consider the series of whole square numbers, 1, 4, 9, 16, 25, 36, 49, 64, etc. By subtracting each of these from the succeeding one, we obtain a new series, which we will name the Series of First Differences, consisting of the numbers, 3, 5, 7, 9, 11, 13, 15, etc. On subtracting from each of these the preceding one, we obtain the Second Differences, which are all constant and equal to 2. We may represent this succession of operations, and their results, in the following table—

A Column of Square Numbers	B First Differences	C Second Differences
1		
	3	
4		$2b$
	5	
a 9		$2d$
	7	
c 16		2
	9	
25		2
36	11	

From the mode in which the last two columns B and C have been formed, it is easy to see that if, for instance, we desire to pass from the number 5 to the succeeding one 7, we must add to the former the constant difference 2; similarly, if from the square number 9 we would pass to the following one 16, we must add to the former the difference 7, which difference is in other words the preceding difference 5, plus the constant difference 2; or again which comes to the same thing, to obtain 16 we have only to add together the three numbers 2, 5, 9, placed obliquely in the direction *ba*. Similarly, we obtain the number 25 by summing up the three numbers placed in the oblique direction *dc*: commencing by the addition 2 + 7, we have the first difference 9 consecutively to 7; adding 16 to the 9 we have the square 25. We see then that the three numbers 2, 5, 9 being given, the whole series of successive square numbers, and that of their first differences likewise, may be obtained by means of simple additions.

Now, to conceive how these operations may be reproduced by a machine, suppose the latter to have three dials, designated as *A*, *B*, *C*, on each of which are traced, say a thousand divisions, by way of example, over which a needle shall pass. The two dials, *C*, *B*, shall have in addition a registering hammer, which is to give a number of strokes equal to that of the divisions indicated by the needle. For each stroke of the registering hammer of the dial *C*, the needle *B* shall advance one division; similarly, the needle *A* shall advance one division for every stroke of the registering hammer of the dial *B*. Such is the general disposition of the mechanism.

This being understood, let us at the beginning of the series of operations we wish to execute, place the needle *C* on the division 2, the needle *B* on the division 5, and the needle *A* on the division 9. Let us allow the hammer of the dial *C* to strike; it will strike twice, and at the same time the needle *B* will pass over two divisions. The latter will then indicate the number 7, which succeeds the number 5 in the column of first differences. If we now permit the hammer of the dial *B* to strike in its turn, it will strike seven times, during which the needle *A* will advance seven divisions; these added to the nine already marked by it, will give the number 16, which is the square number consecutive to 9. If we now recommence these operations, beginning with the needle *C*, which is always to be left on the division 2, we shall perceive that by repeating them indefinitely, we may successively reproduce the series of whole square numbers by means of a very simple mechanism.

8

The theorem on which is based the construction of the machine we have just been describing, is a particular case of the following more general theorem: that if in any polynomial whatever, the highest power of whose variable is m, this same variable be increased by equal degrees; the corresponding values of the polynomial then calculated, and the first, second, third, etc., differences of these be taken (as for the preceding series of squares); the mth differences will all be equal to each other. So that, in order to reproduce the series of values of the polynomial by means of a machine analogous to the one above described, it is sufficient that there be $(m+1)$ dials, having the mutual relations we have indicated. As the differences may be either positive or negative, the machine will have a contrivance for either advancing or retrograding each needle, according as the number to be algebraically added may have the sign *plus* or *minus*.

If from a polynomial we pass to a series having an infinite number of terms, arranged according to the ascending powers of the variable, it would at first appear, that in order to apply the machine to the calculation of the function represented by such a series, the mechanism must include an infinite number of dials, which would in fact render the thing impossible. But in many cases the difficulty will disappear, if we observe that for a great number of functions the series which represent them may be rendered convergent; so that, according to the degree of approximation desired, we may limit ourselves to the calculation of a certain number of terms of the series, neglecting the rest. By this method the question is reduced to the primitive case of a finite polynomial. It is thus that we can calculate the succession of the logarithms of numbers. But since, in this particular instance, the terms which had been originally neglected receive increments in a ratio so continually increasing for equal increments of the variable, that the degree of approximation required would ultimately be affected, it is necessary, at certain intervals, to calculate the value of the function by different methods, then respectively to use the results thus obtained, as data whence to deduce, by means of the machine, the other intermediate values. We see that the machine here performs the office of the third section of calculators mentioned in describing the tables computed by order of the French government, and that the end originally proposed is thus fulfilled by it.

Such is the nature of the first machine which Mr. Babbage conceived. We see that its use is confined to cases where the numbers required are such as can be obtained by means of simple additions or subtractions; that the machine is, so to speak, merely the expression of one particular theorem of analysis; and that, in short, its operations cannot be extended so as to embrace the solution of an infinity of other questions included within the domain of mathematical analysis. It was while contemplating the vast field which yet remained to be traversed, that Mr. Babbage, renouncing his original essays, conceived the plan of another system of mechanism whose operations should themselves possess all the generality of algebraical notation, and which, on this account, he denominates the *Analytical Engine*.

Having now explained the state of the question, it is time for me to develop the principle on which is based the construction of this latter machine. When analysis is employed for the solution of any problem, there are usually two classes of operations to execute: first, the numerical calculation of the various coefficients; and secondly, their distribution in relation to the quantities affected by them. If, for example, we have to obtain the product of two binomials $(a+bx)$ $(m+nx)$, the result will be represented by $am+(an+bm)x+bnx^2$, in which expression we must first calculate *am, an, bm, bn;* then take the sum of *an+bm;* and lastly, respectively distribute the coefficients thus obtained amongst the powers of the variable. In order to reproduce these operations by means of a machine, the latter must therefore possess two distinct sets of powers: first, that of executing numerical calculations; secondly, that of rightly distributing the values so obtained.

But if the human intervention were necessary for directing each of these partial operations, nothing would be gained under the heads of correctness and economy of time; the machine must therefore have the additional requisite of executing by itself all the successive operations required for the solution of a problem proposed to it; when once the primitive *numerical data* for this same problem have been introduced. Therefore, since from the moment that the nature of the calculation to be executed and the problem to be resolved have been indicated to it, the machine is, by its own intrinsic power, of itself to go through all the intermediate operations which lead to the proposed result, it must exclude all methods of trial and guess-work, and can only admit the direct processes of calculation.*

* This must not be understood in too unqualified a manner. The engine is capable, under certain circumstances, of feeling about to discover which of two or more possible contingencies has occurred, and of then shaping its future course accordingly.—Note by Translator.

It is necessarily thus; for the machine is not a thinking being, but simply an automaton which acts according to the laws imposed upon it. This being fundamental, one of the earliest researches its author had to undertake, was that of finding means for effecting the division of one number by another without using the method of guessing indicated by the usual rules of arithmetic. The difficulties of effecting this combination were far from being among the least; but upon it depended the success of every other. Under the impossibility of my here explaining the process through which this end is attained, we must limit ourselves to admitting that the first four operations of arithmetic, that is addition, subtraction, multiplication and division, can be performed in a direct manner through the intervention of the machine. This granted, the machine is thence capable of performing every species of numerical calculation, for all such calculations ultimately resolve themselves into the four operations we have just named. To conceive how the machine can now go through its functions according to the laws laid down, we will begin by giving an idea of the manner in which it materially represents numbers.

Let us conceive a pile or vertical column consisting of an indefinite number of circular discs, all pierced through their centres by a common axis, around which each of them can take an independent rotatory movement. If round the edge of each of these discs are written the ten figures which constitute our numerical alphabet, we may then, by arranging a series of these figures in the same vertical line, express in this manner any number whatever. It is sufficient for this purpose that the first disc represent units, the second tens, the third hundreds, and so on. When two numbers have been written on two distinct columns, thus we may propose to combine them arithmetically with each other, and to obtain the result on a third column. In general, if we have a series of columns consisting of discs, which columns we will designate as V_0, V_1, V_2, V_3, V_4, etc., we may require for instance, to divide the number written on the column V_1 by that on the column V_4, and to obtain the result on the column V_7. To effect this operation, we must impart to the machine two distinct arrangements; through the first it is prepared for executing a *division*, and through the second the columns it is to operate on are indicated to it, and also the column on which the result is to be represented. If this division is to be followed, for example, by the addition of two numbers taken on other columns, the two original arrangements of the machine must be simultaneously altered. If, on the contrary, a series of operations of the same nature is to be gone through, then the first of the original arrangements will remain, and the second alone must be altered. Therefore, the arrangements that may be communicated to the various parts of the machine, may be distinguished into two principal classes:

First, that relative to the *Operations*.

Secondly, that relative to the *Variables*.

By this latter we mean that which indicates the columns to be operated on. As for the operations themselves, they are executed by a special apparatus, which is designated by the name of *mill*, and which itself contains a certain number of columns, similar to those of the Variables. When two numbers are to be combined together, the machine commences by effacing them from the columns where they are written, that is it places *zero** on every disc of the two vertical lines on which the numbers were represented; and it transfers the numbers to the mill. There, the apparatus having been disposed suitably for the required operation, this latter is effected, and, when completed, the result itself is transferred to the column of Variables which shall have been indicated. Thus the mill is that portion of the machine which works, and the columns of Variables constitute that where the results are represented and arranged. After the preceding explanations, we may perceive that all fractional and irrational results will be represented in decimal fractions. Supposing each column to have forty discs, this extension will be sufficient for all degrees of approximation generally required.

It will now be inquired how the machine can of itself, and without having recourse to the hand of man, assume the successive dispositions suited to the operations. The solution of this problem has been taken from Jacquard's apparatus, used for the manufacture of brocaded stuffs, in the following manner—

Two species of threads are usually distinguished in woven stuffs; one is the *warp* or longitudinal thread, the other the *woof* or transverse thread, which is conveyed by the instrument called the shuttle, and which crosses the longitudinal thread or warp. When a brocaded stuff is required, it is necessary in turn to prevent certain threads from crossing the woof, and this according to a succession which is determined by the nature of the design that is to be reproduced. Formerly this process was lengthy and difficult, and it was requisite that the

* Zero is not *always* substituted when a number is transferred to the mill. This is explained further on in the memoir.—Note by Translator.

workman, by attending to the design which he was to copy, should regulate the movements the threads were to take. Thence arose the high price of this description of stuffs, especially if threads of various colours entered into the fabric. To simplify this manufacture, Jacquard devised the plan of connecting each group of threads that were to act together, with a distinct lever belonging exclusively to that group. All these levers terminate in rods, which are united together in one bundle, having usually the form of a parallelepiped with a rectangular base. The rods are cylindrical, and are separated from each other by small intervals. The process of raising the threads is thus resolved into that of moving these various lever-arms in the requisite order. To effect this, a rectangular sheet of pasteboard is taken, somewhat larger in size than a section of the bundle of lever-arms. If this sheet be applied to the base of the bundle, and an advancing motion be then communicated to the pasteboard, this latter will move with it all the rods of the bundle, and consequently the threads that are connected with each of them. But if the pasteboard, instead of being plain, were pierced, with holes corresponding to the extremities of the levers which meet it, then, since each of the levers would pass through the pasteboard during the motion of the latter, they would all remain in their places. We thus see that it is easy so to determine the position of the holes in the pasteboard, that, at any given moment, there shall be a certain number of levers, and consequently of parcels of threads, raised, while the rest remain where they were. Supposing this process is successively repeated according to a law indicated by the pattern to be executed, we perceive that this pattern may be reproduced on the stuff. For this purpose we need merely compose a series of cards according to the law required, and arrange them in suitable order one after the other; then, by causing them to pass over a polygonal beam which is so connected as to turn a new face for every stroke of the shuttle, which face shall then be impelled parallelly to itself against the bundle of lever-arms, the operation of raising the threads will be regularly performed. Thus we see that brocaded tissues may be manufactured with a precision and rapidity formerly difficult to obtain.

Arrangements analogous to those just described have been introduced into the Analytical Engine. It contains two principal species of cards: first, Operation cards, by means of which the parts of the machine are so disposed as to execute any determinate series of operations, such as additions, subtractions, multiplications, and divisions; secondly, cards of the Variables, which indicate to the machine the columns on which the results are to be represented. The cards, when put in motion, successively arrange the various portions of the machine according to the nature of the processes that are to be effected, and the machine at the same time executes these processes by means of the various pieces of mechanism of which it is constituted.

In order more perfectly to conceive the thing, let us select as an example the resolution of two equations of the first degree with two unknown quantities. Let the following be the two equations, in which x and y are the unknown quantities—

$$\begin{cases} mx + ny = d \\ m'x + n'y = d' \end{cases}$$

We deduce $x = \dfrac{dn' - d'n}{n'm - nm'}$, and for y an analogous expression. Let us continue to represent by V_0, V_1, V_2, etc. the different columns which contain the numbers, and let us suppose that the first eight columns have been chosen for expressing on them the numbers represented by m, n, d, m', n', d', n and n', which implies that $V_0 = m$, $V_1 = n$, $V_2 = d$, $V_3 = m'$, $V_4 = n'$, $V_5 = d'$, $V_6 = n$, $V_7 = n'$.

The series of operations commanded by the cards, and the results obtained, may be represented in the facing table.

Since the cards do nothing but indicate in what manner and on what columns the machine shall act, it is clear that we must still, in every particular case, introduce the numerical data for the calculation. Thus, in the example we have selected, we must previously inscribe the numerical values of m, n, d, m', n', d', in the order and on the columns indicated, after which the machine when put in action will give the value of the unknown quantity x for this particular case. To obtain the value of y, another series of operations analogous to the preceding must be performed. But we see that they will be only four in number, since the denominator of the expression for y, excepting the sign, is the same as that for x, and equal to $n'm - nm'$. In the preceding table it will be remarked that the column for operations indicates four successive *multiplications*, two *subtractions*, and one *division*. Therefore, if desired, we need only use three operation cards; to manage which, it is sufficient to introduce into the machine an apparatus which shall, after the first multiplication, for instance, retain the card which relates to this operation, and not allow it to advance so as to be replaced by another one, until after this same operation shall have been four times repeated. In the preceding example we have seen, that to find the value of x we must begin by writing the coefficients m, n, d, m', n', d', upon eight columns, thus repeating n and n' twice. According to the same method, if it were required to calculate y likewise, these coefficients must be written on twelve different columns. But it is possible to simplify this process, and thus to diminish the chances of errors, which chances are greater, the larger the number of the quantities that

Number of the operations	Operation-cards	Cards of the variables		Progress of the operations
	Symbols indicating the nature of the operations	Columns on which operations are to be performed	Columns which receive results of operations	
1	$\times$	$V_2 \times V_4 =$	V_8 . .	$= dn'$
2	$\times$	$V_5 \times V_1 =$	V_9 . .	$= d'n$
3	$\times$	$V_4 \times V_0 =$	V_{10} . .	$= n'm$
4	$\times$	$V_1 \times V_3 =$	V_{11} . .	$= nm'$
5	$-$	$V_8 - V_9 =$	V_{12} . .	$= dn' - d'n$
6	$-$	$V_{10} - V_{11} =$	V_{13} . .	$= n'm - nm'$
7	$\div$	$\dfrac{V_{12}}{V_{13}} =$	V_{14} . .	$= x = \dfrac{dn' - d'n}{n'm - nm'}$

have to be inscribed previous to setting the machine in action. To understand this simplification, we must remember that every number written on a column must, in order to be arithmetically combined with another number, be effaced from the column on which it is, and transferred to the *mill*. Thus, in the example we have discussed, we will take the two coefficients m and n', which are each of them to enter into *two* different products, that is m into mn' and md', n' into mn' and $n'd$. These coefficients will be inscribed on the columns V_0 and V_4. If we commence the series of operations by the product of m into n', these numbers will be effaced from the columns V_0 and V_4, that they may be transferred to the mill, which will multiply them into each other, and will then command the machine to represent the result, say on the column V_6. But as these numbers are each to be used again in another operation, they must again be inscribed somewhere; therefore, while the mill is working out their product, the machine will inscribe them anew on any two columns that may be indicated to it through the cards; and, as in the actual case, there is no reason why they should not resume their former places, we will suppose them again inscribed on V_0 and V_4, whence in short they would not finally disappear, to be reproduced no more, until they should have gone through all the combinations in which they might have to be used.

We see, then, that the whole assemblage of operations requisite for resolving the two above equations of the first degree, may be definitely represented in the table on page 17.

In order to diminish to the utmost the chances of error in inscribing the numerical data of the problem, they are successively placed on one of the columns of the mill; then, by means of cards arranged for this purpose, these same numbers are caused to arrange themselves on the requisite columns, without the operator having to give his attention to it; so that his undivided mind may be applied to the simple inscription of these same numbers.

According to what has now been explained, we see that the collection of columns of Variables may be regarded as a *store* of numbers accumulated there by the mill, and which, obeying the orders transmitted to the machine by means of the cards, pass alternately from the mill to the store, and from the store to the mill, that they may undergo the transformations demanded by the nature of the calculation to be performed.

Hitherto no mention has been made of the *signs* in the results, and the machine would be far from perfect were it incapable of expressing and combining amongst each other positive and negative quantities. To accomplish this end, there is, above every column, both of the mill and of the store, a disc, similar to the discs of which the columns themselves consist. According as the digit on this disc is even or uneven, the number inscribed on the corresponding column below it will be considered as positive or negative. This granted, we may, in the following manner, conceive how the signs can be algebraically combined in the machine. When a number is to be transferred from the store to the mill, and vice versa, it will always be transferred with its sign, which will be effected by means of the cards, as has been explained in what precedes. Let any two numbers then, on which we are to operate arithmetically, be placed in the mill with their respective signs. Suppose that we are first to add them together; the operation-cards will command the addition: if the two numbers be of the same sign, one of the two will be entirely effaced from where it was inscribed, and will go to add itself on the column which contains the other number; the machine will, during this operation, be able, by means of a certain apparatus to prevent any movement in the disc of signs which belongs to the column on which the addition is made, and thus the result

Columns which are described the primitive data	Number of the operations	Cards of the operations		Variable cards			Statement of results
		Number of the operation cards	Nature of each operation	Columns acted on by each operation	Columns that receive the result of each operation	Indication of change of value on any column	
$_0 = m$	1	1	×	$^1V_0 \times {}^1V_4 =$	$^1V_6 \ldots$	$\begin{pmatrix} {}^1V_0 = {}^1V_0 \\ {}^1V_4 = {}^1V_4 \end{pmatrix}$	$^1V_6 = mn'$
$_1 = n$	2	,,	×	$^1V_3 \times {}^1V_1 =$	$^1V_7 \ldots$	$\begin{pmatrix} {}^1V_3 = {}^1V_3 \\ {}^1V_1 = {}^1V_1 \end{pmatrix}$	$^1V_7 = m'n$
$_2 = d$	3	,,	×	$^1V_2 \times {}^1V_4 =$	$^1V_8 \ldots$	$\begin{pmatrix} {}^1V_2 = {}^1V_2 \\ {}^1V_4 = {}^0V_4 \end{pmatrix}$	$^1V_8 = dn'$
$_3 = m'$	4	,,	×	$^1V_5 \times {}^1V_1 =$	$^1V_9 \ldots$	$\begin{pmatrix} {}^1V_5 = {}^1V_5 \\ {}^1V_1 = {}^0V_1 \end{pmatrix}$	$^1V_9 = d'n$
$_4 = n'$	5	,,	×	$^1V_0 \times {}^1V_5 =$	$^1V_{10} \ldots$	$\begin{pmatrix} {}^1V_0 = {}^0V_0 \\ {}^1V_5 = {}^0V_5 \end{pmatrix}$	$^1V_{10} = d'm$
$_5 = d'$	6	,,	×	$^1V_2 \times {}^1V_3 =$	$^1V_{11} \ldots$	$\begin{pmatrix} {}^1V_2 = {}^0V_2 \\ {}^1V_3 = {}^0V_3 \end{pmatrix}$	$^1V_{11} = dm'$
	7	2	−	$^1V_6 - {}^1V_7 =$	$^1V_{12} \ldots$	$\begin{pmatrix} {}^1V_6 = {}^0V_6 \\ {}^1V_7 = {}^0V_7 \end{pmatrix}$	$^1V_{12} = mn' - m'n$
	8	,,	−	$^1V_8 - {}^1V_9 =$	$^1V_{13} \ldots$	$\begin{pmatrix} {}^1V_8 = {}^0V_8 \\ {}^1V_9 = {}^0V_9 \end{pmatrix}$	$^1V_{13} = dn' - d'n$
	9	,,	−	$^1V_{10} - {}^1V_{11} =$	$^1V_{14} \ldots$	$\begin{pmatrix} {}^1V_{10} = {}^0V_{10} \\ {}^1V_{11} = {}^0V_{11} \end{pmatrix}$	$^1V_{14} = d'm - dm'$
	10	3	÷	$^1V_{13} \div {}^1V_{12} =$	$^1V_{15} \ldots$	$\begin{pmatrix} {}^1V_{13} = {}^0V_{13} \\ {}^1V_{12} = {}^1V_{12} \end{pmatrix}$	$^1V_{15} = \dfrac{dn' - d'n}{mn' - m'n} = x$
	11	,,	÷	$^1V_{14} \div {}^1V_{12} =$	$^1V_{16} \ldots$	$\begin{pmatrix} {}^1V_{14} = {}^0V_{14} \\ {}^1V_{12} = {}^0V_{12} \end{pmatrix}$	$^1V_{16} = \dfrac{d'm - dm'}{mn' - m'n} = y$
1	2	3	4	5	6	7	8

will remain with the sign which the two given numbers originally had. When two numbers have two different signs, the addition commanded by the card will be changed into a subtraction through the intervention of mechanisms which are brought into play by this very difference of sign. Since the subtraction can only be effected on the larger of the two numbers, it must be arranged

that the disc of signs of the larger number shall not move while
the smaller of the two numbers is being effaced from its column
and subtracted from the other, whence the result will have the
sign of this latter, just as in fact it ought to be. The combinations
to which algebraical subtraction give rise, are analogous to the
preceding. Let us pass on to multiplication. When two numbers
to be multiplied are of the same sign, the result is positive; if
the signs are different, the product must be negative. In order
that the machine may act conformably to this law, we have but
to conceive that on the column containing the product of the
two given numbers, the digit which indicates the sign of that
product, has been formed by the mutual addition of the two digits
that respectively indicated the signs of the two given numbers;
it is then obvious that if the digits of the signs are both even, or
both odd, their sum will be an even number, and consequently
will express a positive number; but that if, on the contrary, the
two digits of the signs are one even and the other odd, their
sum will be an odd number, and will consequently express a
negative number. In the case of division, instead of adding the
digits of the discs, they must be subtracted one from the other,
which will produce results analogous to the preceding; that is
to say, that if these figures are both even or both uneven, the
remainder of this subtraction will be even; and it will be uneven
in the contrary case. When I speak of mutually adding or subtracting the numbers expressed by the digits of the signs, I
merely mean that one of the sign-discs is made to advance or
retrograde a number of divisions equal to that which is expressed
by the digit on the other sign-disc. We see, then, from the preceding explanation, that it is possible mechanically to combine
the signs of quantities so as to obtain results conformable to
those indicated by algebra.*

The machine is not only capable of executing those numerical
calculations which depend on a given algebraical formula, but it
is also fitted for analytical calculations in which there are one
or several variables to be considered. It must be assumed that
the analytical expression to be operated on can be developed
according to powers of the variable, or according to determinate
functions of this same variable, such as circular functions, for
instance; and similarly for the result that is to be attained.

*Not having had leisure to discuss with Mr. Babbage the manner of introducing
into his machine the combination of algebraical signs, I do not pretend here to
expose the method he uses for this purpose; but I considered that I ought myself
to supply the deficiency, conceiving that this paper would have been imperfect
if I had omitted to point out one means that might be employed for resolving
this essential part of the problem in question.

If we then suppose that above the columns of the store, we have inscribed the powers or the functions of the variable, arranged according to whatever is the prescribed law of development, the coefficients of these several terms may be respectively placed on the corresponding column below each. In this manner we shall have a representation of an analytical development; and, supposing the position of the several terms composing it to be invariable, the problem will be reduced to that of calculating their coefficients according to the laws demanded by the nature of the question. In order to make this more clear, we shall take the following very simple example, in which we are to multiply $(a+bx^1)$ by $(A+B \cos^1 x)$. We shall begin by writing x^0, $x^1 \cos^0 x$, $\cos^1 x$, above the columns V_0, V_1, V_2, V_3; then, since from the form of the two functions to be combined, the terms which are to compose the products will be of the following nature, $x^0 . \cos^0 x$, $x^0 . \cos^1 x$, $x^1 . \cos^0 x$, $x^1 . \cos^1 x$; these will be inscribed above the columns V_4, V_5, V_6, V_7. The coefficients of x^0, x^1, $\cos^0 x$, $\cos^1 x$ being given, they will, by means of the mill, be passed to the columns V_0, V_1, V_2, and V_3. Such are the primitive data of the problem. It is now the business of the machine to work out its solution, that is to find the coefficients which are to be inscribed on V_4, V_5, V_6, V_7. To attain this object, the law of formation of these same coefficients being known, the machine will act through the intervention of the cards, in the manner indicated by the table on page 20.

It will now be perceived that a general application may be made of the principle developed in the preceding example, to every species of process which it may be proposed to effect on series submitted to calculation. It is sufficient that the law of formation of the coefficients be known, and that this law be inscribed on the cards of the machine, which will then of itself execute all the calculations requisite for arriving at the proposed result. If, for instance, a recurring series were proposed, the law of formation of the coefficients being here uniform, the same operations which must be performed for one of them will be repeated for all the others; there will merely be a change in the locality of the operation, that is it will be performed with different columns. Generally, since every analytical expression is susceptible of being expressed in a series ordered according to certain functions of the variable, we perceive that the machine will include all analytical calculations which can be definitively reduced to the formation of coefficients according to certain laws, and to the distribution of these with respect to the variables.

We may deduce the following important consequence from these explanations, viz. that since the cards only indicate the nature of the operations to be performed, and the columns of Variables with which they are to be executed, these cards will themselves possess all the generality of analysis, of which they are in fact merely a translation. We shall now further examine some of the difficulties which the machine must surmount, if its assimilation to analysis is to be complete. There are certain functions which necessarily change in nature when they pass through zero or infinity, or whose values cannot be admitted when they pass these limits. When such cases present themselves

* Columns above which are written are functions of the variable		Coefficients		Cards of the operations		Cards of the variables			
		Given	To be formed	Number of the operation	Nature of the operation	Columns on which operations are to be performed	Columns on which are to be inscribed the results of the operations	Indication of change of value on any column submitted to an operation	Results of the operations
x^0	$^1V_0$	a							
x^1	$^1V_1$	b							
$\cos^0 x$	$^1V_2$	A							
$\cos^1 x$	$^1V_3$	B							
$x^0 \cos^0 x$	$^0V_4$	–	aA	1	×	$^1V_0 \times {}^1V_2 =$	$^1V_4$ . .	$\begin{cases}^1V_0 = {}^1V_0 \\ ^1V_2 = {}^1V_3\end{cases}$	$^1V_4 = aA$ coefficients of $x^0 \cos^0 x$
$x^0 \cos^1 x$	$^0V_5$	–	aB	2	×	$^1V_0 \times {}^1V_3 =$	$^1V_5$ . .	$\begin{cases}^1V_0 = {}^0V_0 \\ ^1V_3 = {}^1V_3\end{cases}$	$^1V_5 = aB$,, ,, $x^0 \cos^1 x$
$x^1 \cos^0 x$	$^0V_6$	–	bA	3	×	$^1V_1 \times {}^1V_2 =$	$^1V_6$ . .	$\begin{cases}^1V_1 = {}^1V_1 \\ ^1V_2 = {}^0V_2\end{cases}$	$^1V_6 = bA$,, ,, $x^1 \cos^0 x$
$x^1 \cos^1 x$	$^0V_7$	–	bB	4	×	$^1V_1 \times {}^1V_3 =$	$^1V_7$ . .	$\begin{cases}^1V_1 = {}^0V_1 \\ ^1V_3 = {}^0V_3\end{cases}$	$^1V_7 = bB$,, ,, $x^1 \cos^1 x$

the machine is able, by means of a bell, to give notice that the passage through zero or infinity is taking place, and it then stops until the attendant has again set it in action for whatever process it may next be desired that it shall perform. If this process has been foreseen, then the machine, instead of ringing, will so dispose itself as to present the new cards which have relation to the operation that is to succeed the passage through zero and infinity. These new cards may follow the first, but may only come into play contingently upon one or other of the two circumstances just mentioned taking place.

* For an explanation of the upper left-hand indices attached to the V's in this and in the preceding table, we must refer the reader to notes appended to the memoir.—Note by Translator.

Let us consider a term of the form ab^n; since the cards are but a translation of the analytical formula, their number in this particular case must be the same, whatever be the value of n; that is to say, whatever be the number of multiplications required for elevating b to the nth power (we are supposing for the moment that n is a whole number). Now, since the exponent n indicates that b is to be multiplied n times by itself, and all these operations are of the same nature, it will be sufficient to employ one single operation-card, viz. that which orders the multiplication.

But when n is given for the particular case to be calculated, it will be further requisite that the machine limit the number of its multiplications according to the given values. The process may be thus arranged. The three numbers a, b and n will be written on as many distinct columns of the store; we shall designate them V_0, V_1, V_2; the result ab^n will place itself on the column V_3. When the number n has been introduced into the machine, a card will order a certain registering-apparatus to mark $(n-1)$, and will at the same time execute the multiplication of b by b. When this is completed, it will be found that the registering-apparatus has effaced a unit, and that it only marks $(n-2)$; while the machine will now again order the number b written on the column V_1, to multiply itself with the product b^2 written on the column V_3, which will give b^3. Another unit is then effaced from the registering-apparatus, and the same processes are continually repeated until it only marks zero. Thus the number b^n will be found inscribed on V_3, when the machine, pursuing its course of operations, will order the product of b^n by a; and the required calculation will have been completed without there being any necessity that the number of operation-cards used should vary with the value of n. If n were negative, the cards, instead of ordering the multiplication of a by b^n, would order its division; this we can easily conceive, since every number being inscribed with its respective sign, is consequently capable of reacting on the nature of the operations to be executed. Finally, if n were fractional, of the form $\frac{p}{q}$, an additional column would be used for the inscription of q, and the machine would bring into action two sets of processes, one for raising b to the power p, the other for extracting the qth root of the number so obtained.

Again, it may be required, for example, to multiply an expression of the form $ax^m + bx^n$ by another Ax^p, $+ Bx^q$, and then to reduce the product to the least number of terms, if any of the indices are equal. The two factors being ordered with respect to

x, the general result of the multiplication would be $Aax^{m+p} + Abx^{n+p} + Bax^{m+q} + Bbx^{n+q}$. Up to this point the process presents no difficulties; but suppose that we have $m = p$ and $n = q$, and that we wish to reduce the two middle terms to a single one $(Ab + Ba)x^{m+q}$. For this purpose, the cards may order $m + q$ and $n + p$ to be transferred into the mill, and there subtracted one from the other; if the remainder is nothing, as would be the case on the present hypothesis, the mill will order other cards to bring to it the coefficients Ab and Ba, that it may add them together and give them in this state as a coefficient for the single term $x^{m+p} = x^{m+q}$.

This example illustrates how the cards are able to reproduce all the operations which intellect performs in order to attain a determinate result, if these operations are themselves capable of being precisely defined.

Let us now examine the following expression:

$$2. \quad \frac{2^2 . 4^2 . 6^2 . 8^2 . 10^2 . . . (2n)^2}{1^2 . 3^2 . 5^2 . 7^2 . 9^2 . . . (2n-1)^2 . (2n+1)^2}$$

which we know becomes equal to the ratio of the circumference to the diameter, when n is infinite. We may require the machine not only to perform the calculation of this fractional expression, but further to give indication as soon as the value becomes identical with that of the ratio of the circumference to the diameter when n is infinite, a case in which the computation would be impossible. Observe that we should thus require of the machine to interpret a result not of itself evident, and that this is not amongst its attributes, since it is no thinking being. Nevertheless, when the case of $n = \infty$ has been foreseen, a card may immediately order the substitution of the value of π (π being the ratio of the circumference to the diameter), without going through the series of calculations indicated. This would merely require that the machine contain a special card, whose office it should be to place the number π in a direct and independent manner on the column indicated to it. And here we should introduce the mention of a third species of cards, which may be called *cards of numbers*. There are certain numbers, such as those expressing the ratio of the circumference to the diameter, the Numbers of Bernoulli, etc., which frequently present themselves in calculations. To avoid the necessity for computing them every time they have to be used, certain cards may be combined specially in order to give these numbers ready made into the mill, whence they afterwards go and place themselves on those columns of the store that are destined for them. Through this means the machine will be susceptible of those simplifications afforded by the use of numerical tables. It would be equally possible to introduce, by means of these cards, the logarithms of numbers; but perhaps it might not be in this case either the shortest or the most appropriate method; for the machine might be able to perform the same calculations by other more expeditious combinations, founded on the rapidity with which it executes the four first operations of arithmetic. To give an idea of this rapidity, we need only mention that Mr. Babbage believes he can, by his engine, form the product of two numbers, each containing twenty figures, in *three minutes*.

Perhaps the immense number of cards required for the solution of any rather complicated problem may appear to be an obstacle; but this does not seem to be the case. There is no limit to the number of cards that can be used. Certain stuffs require for their fabrication not less than *twenty thousand* cards, and we may unquestionably far exceed even this quantity.

Resuming what we have explained concerning the Analytical Engine, we may conclude that it is based on two principles: the first, consisting in the fact that every arithmetical calculation ultimately depends on four principal operations—addition, subtraction, multiplication, and division; the second, in the possibility of reducing every analytical calculation to that of the coefficients for the several terms of a series. If this last principle be true, all the operations of analysis come within the domain of the engine. To take another point of view: the use of cards offers a generality equal to that of algebraical formulae, since such a formula simply indicates the nature and order of the operations requisite for arriving at a certain definite result, and similarly the cards merely command the engine to perform these same operations; but in order that the mechanisms may be able to act to any purpose, the numerical data of the problem must in every particular case be introduced. Thus the same series of cards will serve for all questions whose sameness of nature is such as to require nothing altered excepting the numerical data. In this light the cards are merely a translation of algebraical formulae, or, to express it better, another form of analytical notation.

Since the engine has a mode of acting peculiar to itself, it will in every particular case be necessary to arrange the series of calculations conformably to the means which the machine possesses; for such or such a process which might be very easy for a calculator, may be long and complicated for the engine, and vice versa.

Considered under the most general point of view, the essential object of the machine being able to calculate, according to the laws dictated to it, the values of numerical coefficients which it is then to distribute appropriately on the columns which represent the variables, it follows that the interpretation of formulae and of results is beyond its province, unless indeed this very interpretation be itself susceptible of expression by means of the symbols which the machine employs. Thus, although it is not itself the being that reflects, it may yet be considered as the being which executes the conceptions of intelligence. The cards receive the impress of these conceptions, and transmit to the various trains of mechanism composing the engine the orders necessary for their action. When once the engine shall have been constructed, the difficulty will be reduced to the making out of the cards; but as these are merely the translation of algebraical formulae, it will by means of some simple notations, be easy to consign the execution of them to a workman. Thus the whole intellectual labour will be limited to the preparation of the formulae, which must be adapted for calculation by the engine.

Now, admitting that such an engine can be constructed, it may be inquired: what will be its utility? To recapitulate; it will afford the following advantages: first, rigid accuracy. We know that numerical calculations are generally the stumbling-block to the solution of problems, since errors easily creep into them, and it is by no means always easy to detect these errors. Now the engine, by the very nature of its mode of acting, which requires no human intervention during the course of its operations, presents every species of security under the head of correctness; besides, it carries with it its own check; for at the end of every operation it prints off, not only the results, but likewise the numerical data of the question; so that it is easy to verify whether the question has been correctly proposed. Secondly, economy of time: to convince ourselves of this, we need only recollect that the multiplication of two numbers, consisting each of twenty figures, requires at the very utmost three minutes. Likewise, when a long series of identical computations is to be performed, such as those required for the formation of numerical tables, the machine can be brought into play so as to give several results at the same time, which will greatly abridge the whole amount of the processes. Thirdly, economy of intelligence: a simple arithmetical computation requires to be performed by a person possessing some capacity; and when we pass to more complicated calculations, and wish to use algebraical formulae in particular cases, knowledge must be possessed which pre-supposses preliminary mathematical studies of some extent. Now the engine, from its capability of performing by itself all these purely material operations, spares intellectual labour, which may be more profitably employed. Thus the engine may be considered as a real manufactory of figures, which will lend its aid to those many useful sciences and arts that depend on numbers. Again, who can foresee the consequences of such an invention? In truth, how many precious observations remain practically barren for the progress of the sciences, because there are not powers sufficient for computing the results! And what discouragement does the perspective of a long and arid computation cast into the mind of a man of genius, who demands time exclusively for meditation, and beholds it snatched from him by the material routine of operations! Yet it is by the laborious route of analysis that he must reach the truth; but he cannot pursue this unless guided by numbers; for without numbers it is not given us to raise the veil which envelopes the mysteries of nature. Thus the idea of constructing an apparatus capable of aiding human weakness in such researches, is a conception which, being realized, would mark a glorious epoch in the history of the sciences. The plans have been arranged for all the various parts, and for all the wheel-work, which compose this immense apparatus, and their action studied; but all these have not yet been fully combined together in the drawings* and mechanical notation†. The confidence which the genius of Mr. Babbage must inspire, affords legitimate ground for hope that this enterprise will be crowned with success; and while we render homage to the intelligence which directs it, let us breathe aspirations for the accomplishment of such an undertaking.

* This sentence has been slightly altered in the translation in order to express more exactly the present state of the engine.—Note by Translator.

† The notation here alluded to is a most interesting and important subject, and would have well deserved a separate and detailed note upon it, amongst those appended to the Memoir. It has, however, been impossible, within the space allotted, even to touch upon so wide a field.—Note by Translator.

INTELLIGENT MACHINERY

The great English mathematician Alan Turing's brilliant essay, 'Can a Machine Think?' (included in Butterworth's *Key Papers: Brain Physiology and Psychology*) is essential reading for students interested in the background philosophy of cybernetics. Turing's contribution to the development of the modern computer is as significant as any of his generation, and his tragic death at an early age left a serious gap in the ranks of the British computer pioneers. For a considerable time Turing worked at the National Physical Laboratory in Teddington where he was principally responsible for the programme to develop the (for the time) huge and ultra-fast Computer, ACE, parts of which now repose in the Science Museum in London. While at the NPL, Turing wrote several technical reports, a number of which have for one reason or another not been published in scientific journals. Here is one such, a lucid speculative piece on intelligent machinery which forces home a number of points — not the least of which is that scientific writing may be of the highest calibre and may yet remain eminently readable.

Further Reading

Good, I. J. (1965). 'Speculations concerning the first ultra-intelligent machine'. *Adv. Comput.* Vol. 6. 31-88
Turing, Sarah (1959). *Alan Turing*. Cambridge, Heffer.

INTELLIGENT MACHINERY

A. M. Turing

I propose to investigate the question as to whether it is possible for machinery to show intelligent behaviour. It is usually assumed without argument that it is not possible. Common catch phrases such as 'acting like a machine', 'purely mechanical behaviour' reveal this common attitude. It is not difficult to see why such an attitude should have arisen. Some of the reasons are

(a) An unwillingness to admit the possibility that mankind can have any rivals in intellectual power. This occurs as much amongst intellectual people as amongst others: they have more to lose. Those who admit the possibility all agree that its realization would be very disagreeable. The same situation arises in connection with the possibility of our being superseded by some other animal species. This is almost as disagreeable and its theoretical possibility is indisputable.

(b) A religious belief that any attempt to construct such machines is a sort of Promethean irreverence.

(c) The very limited character of the machinery which has been used until recent times (e.g. up to 1940). This encouraged the belief that machinery was necessarily limited to extremely straightforward, possibly even to repetitive, jobs. This attitude is very well expressed by Dorothy Sayers (*The Mind of the Maker*, p. 46) '...which imagines that God, having created his Universe, has now screwed the cap on His pen, put His feet on the mantelpiece and left the work to get on with itself.' This, however, rather comes into St. Augustine's category of figures of speech or enigmatic sayings framed from things which do not exist at all. We simply do not know of any creation which goes on creating itself in variety when the creator has withdrawn from it. The idea is that God simply created a vast machine and has left it working until it runs down from lack of fuel. This is another of those obscure analogies, since we have no experience of machines that produce variety of their own accord; the nature of a machine is to do the same thing over and over again so long as it keeps going.'

(d) Recently the theorem of Godel and related results (Godel[1], Church[2], Turing[3]) have shown that if one tries to use machines for such purposes as determining the truth or falsity of mathematical theorems and one is not willing to tolerate an occasional wrong result, then any given machine will in some cases be unable to give an answer at all. On the other hand the human intelligence seems to be able to find methods of ever-increasing power for dealing with such problems 'transcending' the methods available to machines.

(e) In so far as a machine can show intelligence this is to be regarded as nothing but a reflection of the intelligence of its creator.

I. REFUTATION OF SOME OBJECTIONS

In this section I propose to outline reasons why we do not need to be influenced by the above described objections. The objections (a) and (b), being purely emotional, do not really need to be refuted. If one feels it necessary to refute them, there is little to be said that could hope to prevail, though the actual production of the machines would probably have some effect. In so far then as we are influenced by such arguments we are bound to be left feeling rather uneasy about the whole project, at any rate for the present. These arguments cannot be wholly ignored, because the idea of 'intelligence' is itself emotional rather than mathematical.

The objection (c) in its crudest form is refuted at once by the actual existence of machinery (ENIAC etc.) which can go on through immense numbers (e.g. $10^{60,000}$ about for ACE) of operations without repetition, assuming no breakdown. The more subtle forms of this objection will be considered at length in sections 11 and 12.

The argument from Godel's and other theorems (objection (d)) rests essentially on the condition that the machine must not make mistakes. But this is not a requirement for intelligence. It is related that the infant Gauss was asked at school to do the addition $15+18+21+\ldots+54$ (or something of the kind) and that he immediately wrote down 483, presumably having calculated it as $(15+54)(54-12)/2.3$. One can imagine circumstances where a foolish master told the child that he ought instead to have added 18 to 15 obtaining 33, then added 21 etc. From some points of view this would be a 'mistake', in spite of the obvious intelligence involved. One can also imagine a situation where the children were given a number of additions to do, of which the first five were all arithmetic progressions, but the sixth was say $23+34+45+\ldots+100+112+122+\ldots+199$. Gauss might have given the answer to this as if it were an arithmetic progression, not having noticed that the ninth term was 112 instead of 111. This would be a definite mistake, which the less intelligent children would not have been likely to make.

The view (d) that intelligence in machinery is merely a reflection of that of its creator is rather similar to the view that the credit for the discoveries of a pupil should be given to his teacher. In such a case the teacher would be pleased with the success of his methods of education, but would not claim the results themselves unless he had actually communicated them to his pupil. He would certainly have envisaged in very broad outline the sort of thing his pupil might be expected to do, but would not expect to foresee any sort of detail. It is already possible to produce machines where this sort of situation arises in a small degree. One can produce 'paper machines' for playing chess. Playing against such a machine gives a definite feeling that one is pitting one's wits against something alive.

These views will all be developed more completely below.

2. VARIETIES OF MACHINERY

It will not be possible to discuss possible means of producing intelligent machinery without introducing a number of technical terms to describe different kinds of existent machinery.

'Discrete' and 'Continuous' Machinery

We may call a machine 'discrete' when it is natural to describe its possible states as a discrete set, the motion of the machine occurring by jumping from one state to another. The states of 'continuous' machinery on the other hand form a continuous manifold, and the behaviour of the machine is described by a curve on this manifold. All machinery can be regarded as continuous, but when it is possible to regard it as discrete it is usually best to do so. The states of discrete machinery will be described as 'configurations'.

'Controlling' and 'Active' Machinery

Machinery may be described as 'controlling' if it only deals with information. In practice this condition is much the same as saying that the magnitude of the machine's effects may be as small as we please, so long as we do not introduce confusion through Brownian movement etc. 'Active' machinery is intended to produce some definite physical effect.

Examples

A Bulldozer is	Continuous Active
A Telephone is	Continuous Controlling
A Brunsviga is	Discrete Controlling
A Brain is probably Continuous Controlling, but is very similar to much discrete machinery
The ENIAC, ACE etc.	Discrete Controlling
A Differential Analyser	Continuous Controlling

We shall mainly be concerned with discrete controlling machinery. As we have mentioned brains very nearly fall into this class, and there seems every reason to believe that they could have been made to fall genuinely into it without any change in their essential properties. However, the property of being 'discrete' is only an advantage for the theoretical investigator, and serves no evolutionary purpose, so we could not expect Nature to assist us by producing truly 'discrete' brains.

Given any discrete machine the first thing we wish to find out about it is the number of states (configurations) it can have. This number may be infinite (but enumerable) in which case we say that the machine has infinite memory (or storage) capacity. If the machine has a finite number N of possible states, then we say that it has a memory capacity of (or equivalent to) $\log_2 N$ binary digits. According to this definition we have the following table of capacities, very roughly

Brunsviga	90
ENIAC without cards and with fixed programme	600
ENIAC with cards	
ACE as proposed	60,000
Manchester machine (as actually working 8/7/48)	1,100

The memory capacity of a machine more than anything else determines the complexity of its possible behaviour.

The behaviour of a discrete machine is completely described when we are given the state (configuration) of the machine as a function of the immediately preceding state and the relevant external data.

LOGICAL COMPUTING MACHINES (L.C.M.s)

In Turing[3] a certain type of discrete machine was described. It had an infinite memory capacity obtained in the form of an infinite tape marked out into squares on each of which a symbol could be printed. At any moment there is one symbol in the machine; it is called the scanned symbol. The machine can alter the scanned symbol and its behaviour is in part determined by that symbol, but the symbols on the tape elsewhere do not affect the behaviour of the machine. However, the tape can be moved back and forth through the machine, this being one of the elementary operations of the machine. Any symbol on the tape may therefore eventually have an innings.

These machines will here be called 'Logical Computing Machines'. They are chiefly of interest when we wish to consider what a machine could in principle be designed to do, when we are willing to allow it both unlimited time and unlimited storage capacity.

Universal Logical Computing Machines

It is possible to describe L.C.M.s in a very standard way, and to put the description into a form which can be 'understood' (i.e. applied by) a special machine. In particular it is possible to design a 'universal machine' which is an L.C.M. such that if the standard description of some other L.C.M. is imposed on the otherwise blank tape from outside, and the (universal) machine then set going it will carry out the operations of the particular machine whose description it was given. For details the reader must refer to Turing[3].

The importance of the universal machine is clear. We do not need to have an infinity of different machines doing different jobs. A single one will suffice. The engineering problem of producing various machines for various jobs is replaced by the office work of 'programming' the universal machine to do these jobs.

It is found in practice that L.C.M.s can do anything that could be described as 'rule of thumb' or 'purely mechanical'. This is sufficiently well established that it is now agreed amongst logicians that 'calculable by means of an L.C.M.' is the correct accurate rendering of such phrases. There are several mathematically equivalent but superficially very different renderings.

PRACTICAL COMPUTING MACHINES (P.C.M.s)

Although the operations which can be performed by L.C.M.s include every rule of thumb process, the number of steps involved tends to be enormous. This is mainly due to the arrangement of the memory along the tape. Two facts which need to be used together may be stored very far apart on the tape. There is also rather little encouragement, when dealing with these machines, to condense the stored expressions at all. For instance, the number of symbols required in order to express a number in Arabic form (e.g. 149056) cannot be given any definite bound, any more than if the numbers are expressed in the 'simplified Roman' form (IIIII...I, with 149056 occurrences of I). As the simplified Roman system obeys very much simpler laws one uses it instead of the Arabic system.

In practice, however, one *can* assign finite bounds to the numbers that one will deal with. For instance, we can assign a bound to the number of steps that we will admit in a calculation performed with a real machine in the following sort of way. Suppose that the storage system depends on charging condensers of capacity $C = 1\mu F$, and that we use two states of charging, $E = 100$ volts and $-E = -100$ volts. When we wish to use the information carried by the condenser we have to observe its voltage. Owing to thermal agitation the voltage observed will always be slightly wrong, and the probability of an error between V and $V - dV$ volts is

$$\frac{2kT}{\pi C} e^{-\frac{1}{2} \frac{V^2 C}{kT}} V dV$$

where k is Boltzmann's constant. Taking the values suggested we find that the probability of reading the sign of the voltage wrong is about $10^{-1.2} \times 10^{16}$. If then a job took more than $10^{10^{17}}$ steps we should be virtually certain of getting the wrong answer, and we may therefore restrict ourselves to jobs with fewer steps. Even a bound of this order might have useful simplifying effects. More practical bounds are obtained by assuming that a light wave must travel at least 1 cm between steps, (this would only be false with a very small machine) and that we could not wait more than 100 years for an answer. This would give a limit of 10^{20}

steps. The storage capacity will probably have a rather similar bound, so that we could use sequences of 20 decimal digits for describing the position in which a given piece of data was to be found, and this would be a really valuable possibility.

Machines of the type generally known as 'Automatic Digital Computing Machines' often make great use of this possibility. They also usually put a great deal of their stored information in a form very different from the tape form. By means of a system rather reminiscent of a telephone exchange it is made possible to obtain a piece of information almost immediately by 'dialling' the position of this information in the store. The delay may be only a few microseconds with some systems. Such machines will be described as 'Practical computing machines'.

UNIVERSAL PRACTICAL COMPUTING MACHINES

Nearly all of the P.C.M.s now under construction have the essential properties of the 'Universal Logical Computing' machines mentioned earlier. In practice, given any job which could have been done on an L.C.M. one can also do it on one of these digital computers. I do not mean that we can design a digital computer to do it, but that we stick to one, say the ACE, and that we can do any required job of the type mentioned on it, by suitable programming. The programming is pure paper work. It naturally occurs to one to ask whether, e.g., the ACE would be truly universal if its memory capacity were infinitely extended. I have investigated this question, and the answer appears to be as follows, though I have not proved any formal mathematical theorem about it. As has been explained, the ACE at present uses finite sequences of digits to describe positions in its memory: they are (Sept. 1947) actually sequences of 9 binary digits. The ACE also works largely for other purposes with sequences of 32 binary digits. If the memory were extended e.g., to 1,000 times its present capacity it would be natural to arrange the memory in blocks of nearly the maximum capacity which can be handled with the 9 digits, and from time to time to switch from block to block. A relatively small part would never be switched. This would contain some of the more fundamental instruction tables and those concerned with switching. This part might be called the 'central part'. One would then need to have a number which described which block was in action at any moment. This number might however be as large as one pleased. Eventually the

33

point would be reached where it could not be stored in a word (32 digits), or even in the central part. One would then have to set aside a block for storing the number, or even a sequence of blocks, say blocks 1, 2, ... n. We should then have to store n, and in theory it would be of indefinite size. This sort of process can be extended in all sorts of ways, but we shall always be left with a positive integer which is of indefinite size and which needs to be stored somewhere, and there seems to be no way out of the difficulty but to introduce a 'tape'. But once this has been done, and since we are only trying to prove a theoretical result one might as well, whilst proving the theorem, ignore all the other forms of storage. One will in fact have a U.L.C.M. with some complications. This, in effect, means that one will not be able to prove any result of the required kind which gives any intellectual satisfaction.

Paper Machines

It is possible to produce the effect of a computing machine by writing down a set of rules of procedure and asking a man to carry them out. Such a combination of a man with written instructions will be called a 'Paper Machine'. A man provided with paper, pencil and rubber, and subject to strict discipline is in effect a universal machine. The expression 'paper machine' will often be used below.

3. PARTIALLY RANDOM AND APPARENTLY PARTIALLY RANDOM MACHINES

It is possible to modify the above described types of discrete machines by allowing several alternative operations to be applied at some points, the alternatives to be chosen by a random process. Such a machine will be described as 'partially random'. If we wish to say definitely that a machine is not of this kind we will describe it as 'determined'. Sometimes a machine may be, strictly speaking, determined but appear superficially as if it were partially random. This would occur if for instance, the digits of the number π were used to determine the choices of a partially random machine, where previously a dice thrower or electronic equivalent had been used. These machines are known as apparently partially random.

4. UNORGANIZED MACHINES

So far we have been considering machines which are designed for a definite purpose (though the universal machines are in a sense an exception). We might instead consider what happens when we make up a machine in a comparatively unsystemic way from some kind of standard components. We could consider some particular machine of this nature and find out what sort of things it is likely to do. Machines which are largely random in their construction in this way will be called 'unorganized machines'. This does not pretend to be an accurate term. It is conceivable that the same machine might be regarded by one man as organized and by another as unorganized.

A typical example of an unorganized machine would be as follows: The machine is made up from a rather large number N of similar units. Each unit has two input terminals, and has an output terminal which can be connected to the input terminals of (0 or more) other units. We may imagine that for each integer r, $1 \leqslant r \leqslant \therefore$ two numbers $i(r)$ and $j(r)$ are chosen at random from $1 \ldots N$ and that we connect the inputs of unit r to the outputs of units $i(r)$ and $j(r)$. All of the units are connected to a central synchronizing unit from which synchronizing pulses are emitted at more or less equal intervals of time. The times when these pulses arrive will be called 'moments'. Each unit is capable of having two states at each moment. These states may be called 0 and 1. The state is determined by the rule that the states of the units from which the input leads come are to be taken at the previous moment multiplied together and the result subtracted from 1. An unorganized machine of this character is shown in the diagram below.

r	$i(r)$	$j(r)$
1	3	2
2	3	5
3	4	5
4	3	4
5	2	5

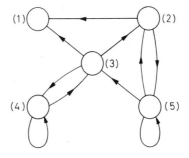

A sequence of possible consecutive conditions for the whole machine is:

1	1	1	0	0	1	0
2	1	1	1	0	1	0
3	0	1	1	1	1	1
4	0	1	0	1	0	1
5	1	0	1	0	1	0

The behaviour of a machine with so few units is naturally very trivial. However, machines of this character can behave in a very complicated manner when the number of units is large. We may call these A-type unorganized machines. Thus the machine in the diagram is an A-type unorganized machine of 5 units. The motion of an A-type machine with N units is of course eventually periodic, as is any determined machine with finite memory capacity. The period cannot exceed 2^N moments, nor can the length of time before the periodic motion begins. In the example above the period is 2 moments and there are 3 moments before the periodic motion begins 2^N is 32.

The A-type unorganized machines are of interest as being about the simplest model of a nervous system with a random arrangement of neurons. It would therefore be of very great interest to find out something about their behaviour. A second type of unorganized machine will now be described, not because it is of any great intrinsic importance, but because it will be useful later for illustrative purposes. Let us denote the circuit

by

as an abbreviation. Then for each A-type unorganized machine we can construct another machine by replacing each connection →— in it by →□→ The resulting machines will be called B-type unorganized machines. It may be said that the B-type machines are all A-type. To this I would reply that the above definitions if correctly (but drily!) set out would take the form of describing the probability of an A- (or B-)type machine belonging to a given set; it is not merely a definition of which are the A-type machines and which are the B-type machines. If one chooses an A-type machine, with a given number of units, at random it will be extremely unlikely that one will get a B-type machine.

It is easily seen that the connection $\longrightarrow\square\longrightarrow$ can have three conditions. It may (i) pass all signals through with interchange of 0 and 1, or (ii) it may convert all signals into 1, or again (iii) it may act as in (i) and (ii) in alternate moments. (Alternative (iii) has two sub-cases). Which of these cases applies depends on the initial conditions. There is a delay of two moments in going through $\longrightarrow\square\longrightarrow$

5. INTERFERENCE WITH MACHINERY.
MODIFIABLE AND SELF-MODIFYING MACHINERY

The types of machine that we have considered so far are mainly ones that are allowed to continue in their own way for indefinite periods without interference from outside. The universal machines were an exception to this, in that from time to time one might change the description of the machine which is being initiated. We shall now consider machines in which such interference is the rule rather than the exception.

We may distinguish two kinds of interference. There is the extreme form in which parts of the machine are removed and replaced by others. This may be described as 'screwdriver interference'. At the other end of the scale is 'paper interference', which consists in the mere communication of information to the machine, which alters its behaviour. In view of the properties of the universal machine we do not need to consider the difference between these two kinds of machine as being so very radical after all. Paper interference when applied to the universal machine can be as useful as screwdriver interference.

We shall mainly be interested in paper interference. Since screwdriver interference can produce a completely new machine without difficulty there is rather little to be said about it. In future, 'interference' will normally mean 'paper interference'.

When it is possible to alter the behaviour of a machine very radically, we may speak of the machine as being 'modifiable'. This is a relative term. One machine may be spoken of as being more modifiable than another.

One may also sometimes speak of a machine modifying itself, or of a machine changing its own instructions. This is really a nonsensical form of phraseology, but is convenient. Of course, according to our conventions the 'machine' is completely described by the relation between its possible configurations at consecutive moments. It is an abstraction which by the form of its definition cannot change in time. If we consider the machine as starting in a particular configuration however, we may be tempted to ignore those configurations which cannot be reached without interference from it. If we do this we should get a 'successor relation' for the configurations with different properties from the original one and so a different 'machine'.

If we now consider interference, we should say that each time interference occurs the machine is probably changed. It is in this sense that interference 'modifies' a machine. The sense in which a machine can modify itself is even more remote. We may if we wish divide the operations of the machine into two classes, normal and self-modifying operations. So long as only normal operations are performed we regard the machine as unaltered. Clearly the idea of 'self-modification' will not be of much interest except where the division of operations into the two classes is very carefully made. The sort of case I have in mind is a computing machine like the ACE where large parts of the storage are normally occupied in holding instruction tables. (Instruction tables are the equivalent in U.P.C.M.s of descriptions of machines in U.L.C.M.s.) Whenever the content of this storage was altered by the internal operations of the machine, one would naturally speak of the machine 'modifying itself'.

6. MAN AS A MACHINE

A great positive reason for believing in the possibility of making thinking machinery is the fact that it is possible to make machinery to imitate any small part of a man. That the microphone does this for the ear, and the television camera for the eye, are commonplaces. One can also produce remote controlled robots whose limbs balance the body with the aid of servo-mechanisms. Here we are chiefly interested in the nervous system. We could produce fairly accurate electrical models to copy the behaviour of nerves, but there seems very little point in doing so. It would be rather like putting a lot of work into cars which walked on legs instead of continuing to use wheels. The electrical circuits which are used in electronic computing machinery seem to have the essential properties of nerves. They are able to transmit information from place to place, and also to store it. Certainly the nerve has many advantages. It is extremely compact, does not wear out (probably for hundreds of years if kept in a suitable medium!) and has a very low energy consumption. Against these advantages the electronic circuits have only one counter attraction, that of speed. This advantage is however on such a scale that it may possibly outweigh the advantages of the nerve.

One way of setting about our task of building a 'thinking machine' would be to take a man as a whole and to try to replace all the parts of him by machinery. He would include television cameras, microphones, loudspeakers, wheels and 'handling servo-mechanisms' as well as some sort of 'electronic brain'. This would of course be a tremendous undertaking. The object if produced by present techniques would be of immense size, even if the 'brain' part were stationary and controlled the body from a distance. In order that the machine should have a chance of finding things out for itself it should be allowed to roam the countryside, and the danger to the ordinary citizen would be serious. Moreover, even when the facilities mentioned above were provided, the creature would still have no contact with food, sex, sport and many other things of interest to the human being. Thus although this method is probably the 'sure' way of producing a thinking machine it seems to be altogether too slow and impracticable.

Instead we propose to try and see what can be done with a 'brain' which is more or less without a body, providing at most, organs of sight, speech and hearing. We are then faced with the problem of finding suitable branches of thought for the machine to exercise its powers in. The following fields appear to me to have advantages:

(a) Various games e.g., chess, noughts and crosses, bridge, poker.
(b) The learning of languages.
(c) Translation of languages.
(d) Cryptography.
(e) Mathematics.

of these (a), (d), and to a lesser extent (c) and (e) are good in that they require little contact with the outside world. For instance, in order that the machine should be able to play chess its only organs need be 'eyes' capable of distinguishing the various positions on a specially made board, and means for announcing its own moves. Mathematics should preferably be restricted to branches where diagrams are not much used. Of the above possible fields the learning of languages would be the most impressive, since it is the most human of these activities. This field seems however to depend rather too much on sense organs and locomotion to be feasible.

The field of cryptography will perhaps be the most rewarding. There is a remarkably close parallel between the problems of the physicist and those of the cryptographer. The system on which a message is enciphered corresponds to the laws of the universe, the intercepted messages to the evidence available, the keys for a day or a message to important constants which have to be determined. The correspondence is very close, but the subject matter of cryptography is very easily dealt with by discrete machinery, physics not so easily.

7. EDUCATION OF MACHINERY

Although we have abandoned the plan to make a 'whole man', we should be wise to sometimes compare the circumstances of our machine with those of a man. It would be quite unfair to expect a machine straight from the factory to compete on equal terms with a university graduate. The graduate has had contact with human beings for 20 years or more. This contact has throughout that period been modifying his behaviour pattern. His teachers have been intentionally trying to modify it. At the end of the period a large number of standard routines will have been superimposed on the original pattern of his brain. These routines will be known to the community as a whole. He is then in a position to try out new combinations of these routines, to make slight variations on them, and to apply them in new ways.

We may say then that in so far as a man is a machine he is one that is subject to very much interference. In fact, interference will be the rule rather than the exception. He is in frequent communication with other men, and is continually receiving visual and other stimuli which themselves constitute a form of interference. It will only be when the man is 'concentrating' with a view to eliminating these stimuli or 'distractions' that he approximates a machine without interference.

We are chiefly interested in machines with comparatively little interference, for reasons given in the last section, but it is important to remember that although a man when concentrating may behave like a machine without interference, his behaviour when concentrating is largely determined by the way he has been conditioned by previous interference.

If we are trying to produce an intelligent machine, and are following the human model as closely as we can, we should begin with a machine with very little capacity to carry out elaborate operations or to react in a disciplined manner to orders (taking the form of interference). Then by applying appropriate interference, mimicking education, we should hope to modify the machine until it could be relied on to produce definite reactions to certain commands. This would be the beginning of the process. I will not attempt to follow it further now.

8. ORGANIZING UNORGANIZED MACHINERY

Many unorganized machines have configurations such that if once that configuration is reached, and if the interference thereafter is appropriately restricted, the machine behaves as one organized for some definite purpose. For instance, the B-type machine shown below was chosen at random

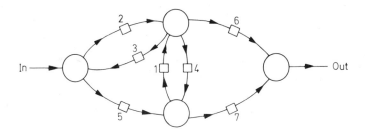

If the connections numbered 1, 3, 6, 4, are in condition (b) initially and connections 2, 5, 7 are in condition (a), then the machine may be considered to be one for the purpose of passing on signals with a delay of 4 moments. This is a particular case of a very general property of B-type machines (and many other types) namely, that with suitable initial conditions they will do any required job, given sufficient time and provided the number of units is sufficient. In particular, with a B-type unorganized machine with sufficient units one can find initial conditions which will make it into a universal machine with a given storage capacity. (A formal proof to this effect might be of some interest, or even a demonstration of it starting with a particular un-

41

organized B-type machine, but I am not giving it as it lies rather too far outside the main argument.)

With these B-type machines the possibility of interference which could set in appropriate initial conditions has not been arranged for. It is, however, not difficult to think of appropriate methods by which this could be done. For instance, instead of the connection

one might use

Here *A*, *B* are interfering inputs, normally giving the signal '1'. By supplying appropriate other signals at *A*, *B* we can get the connection into condition (*a*) or (*b*) or either form of (*c*), as desired. However, this requires two special interfering inputs for each connection.

We shall be mainly interested in cases where there are only quite few independent inputs altogether, so that all the interference which sets up the 'initial conditions' of the machine has to be provided through one or two inputs. The process of setting up these initial conditions so that the machine will carry out some particular useful task may be called 'organizing the machine'. 'Organizing' is thus a form of 'modification'.

9. THE CORTEX AS AN UNORGANIZED MACHINE

Many parts of a man's brain are definite nerve circuits required for quite definite purposes. Examples of these are the 'centres' which control respiration, sneezing, following moving objects with the eyes etc: all the reflexes proper (not 'conditioned') are due to the activities of these definite structures in the brain. Likewise the apparatus for the more elementary analysis of shapes and sounds probably comes into this category. But the more intellectual activities of the brain are too varied to be managed on this basis. The difference between the languages spoken on the two sides of the channel is not due to differences in development of the French-speaking and English-speaking parts of the brain. It is due to the linguistic parts having been subjected to different training. We believe then that there are large parts of the brain, chiefly in the cortex, whose function is largely indeterminate. In the infant these parts do not have much effect: the effect they have is unco-ordinated. In the adult they have great and purposive effect: the form of this effect depends on the training in childhood. A large remnant of the random behaviour of infancy remains in the adult.

All of this suggests that the cortex of the infant is an un-organized machine, which can be organized by suitable inter-fering training. The organizing might result in the modification of the machine into a universal machine or something like it. This would mean that the adult will obey orders given in appro-priate language, even if they were very complicated; he would have no common sense, and would obey the most ridiculous orders unflinchingly. When all his orders had been fulfilled he would sink into a comatose state or perhaps obey some standing order, such as eating. Creatures not unlike this can really be found, but most people behave quite differently under many circumstances. However, the resemblance to a universal machine is still very great, and suggests to us that the step from the unorganized infant to a universal machine is one which should be understood. When this has been mastered we shall be in a far better position to consider how the organizing process might have been modified to produce a more normal type of mind.

This picture of the cortex as an unorganized machine is very satisfactory from the point of view of evolution and genetics. It clearly would not require any very complex system of genes to produce something like the A or B-type unorganized machine. In fact, this should be much easier than the production of such things as the respiratory centre. This might suggest that intelli-gent races could be produced comparatively easily. I think this is wrong because the possession of a human cortex (say) would be virtually useless if no attempt was made to organize it. Thus, if a wolf by a mutation acquired a human cortex there is little reason to believe that he would have any selective advantage. If, however, the mutation occurred in a milieu where speech had developed, (parrot-like wolves), and if the mutation by chance had well permeated their community, then some selective ad-vantage might be felt. It would then be possible to pass infor-mation on from generation to generation. However, this is all rather speculative.

10. EXPERIMENTS IN ORGANIZING.
PLEASURE-PAIN SYSTEMS

It is interesting to experiment with unorganized machines admitting definite types of interference and trying to organize them, e.g. to modify them into universal machines.

The organization of a machine into a universal machine would be most impressive if the arrangements of interference involve very few inputs. The training of the human child depends largely on a system of rewards and punishments, and this suggests that it ought to be possible to carry through the organizing with only two interfering inputs, one for 'pleasure' or 'reward' (R) and the other for 'pain' or 'punishment' (P). One can devise a large number of such 'pleasure-pain' systems. I will use this term to mean an unorganized machine of the following general character:

The configurations of the machine are described by two expressions, which we may call the character-expression and the situation-expression. The character and situation at any moment, together with the input signals, determine the character and situation at the next moment. The character may be subject to some random variation. Pleasure interference has a tendency to fix the character i.e., towards preventing it changing, whereas pain stimuli tend to disrupt the character, causing features which had become fixed to change, or to become again subject to random variation.

This definition is probably too vague and general to be very helpful. The idea is that when the 'character' changes we like to think of it as a change in the machine, but the 'situation' is merely the configuration of the machine described by the character. It is intended that pain stimuli occur when the machine's behaviour is wrong, pleasure stimuli when it is particularly right. With appropriate stimuli on these lines, judiciously operated by the 'teacher' one may hope that the 'character' will converge towards the one desired, i.e. that wrong behaviour will tend to become rare

I have investigated a particular type of pleasure-pain system, which I will now describe.

11. THE P-TYPE UNORGANIZED MACHINE

The P-type machine may be regarded as an L.C.M. without a tape, and whose description is largely incomplete. When a configuration is reached for which the action is undetermined a random choice for the missing data is made and the appropriate entry is made in the description, tentatively, and is applied. When a pain stimulus occurs all tentative entries are cancelled, and when a pleasure stimulus occurs they are all made permanent.

Specifically. The situation is a number $s = 1, 2, ..., N$ and corresponds to the configuration of the incomplete machine. The character is a table of N entries showing the behaviour of the machine in each situation. Each entry has to say something about the next situation and about what action the machine has to take. The action part may be either:

(a) To do some externally visible act A_1 or $A_2 ... A_K$
(b) To set one of the memory units $M_1 ... M$ either into the '1' condition or into the '0' condition.

The next situation is always the remainder either of $2s$ or of $2s + 1$ on division by N. These may be called alternatives 0 and 1. Which alternative applies may be determined by either:

(i) One of the memory units
(ii) A sense stimulus
(iii) The pleasure-pain arrangements

In each situation it is determined which of these applies when the machine is made, i.e. interference cannot alter which of the three cases applies. Also in cases (i) and (ii) interference can have no effect. In case (iii) the entry in the character table may be either U ('uncertain'), or $T0$ (tentative 0), $T1$, $D0$ (definite 0) or $D1$. When the entry in the character for the current situation is U, then the alternative is chosen at random, and the entry in the character is changed to $T0$ or $T1$ according as 0 or 1 was chosen. If the character entry was $T0$ or $D0$, then the alternative is 0 and if it is $T1$ or $D1$, then the alternative is 1. The changes in character include the above-mentioned change from U to $T0$ or $T1$, and a change of every T to D when a pleasure stimulus occurs, changes of $T0$ and $T1$ to U when a pain stimulus occurs.

We may imagine the memory units essentially as 'trigger circuits' or switches. The sense stimuli are means by which the teacher communicates 'unemotionally' to the machine, i.e. otherwise than by pleasure and pain stimuli. There are a finite number S of sense stimulus lines, and each always carries either the signal 0 or 1.

A small P-type machine is described in the table below

1	P	A	
2	P	B	$M1 = 1$
3	P	B	
4	$S1$	A	$M1 = 0$
5	$M1$	C	

In this machine there is only one memory unit $M1$ and one sense line $S1$. Its behaviour can be described by giving the successive situations together with the actions of the teacher: the latter consist of the values of $S1$ and the rewards and punishments. At any moment the 'character' consists of the above table with each 'P' replaced by either U, $T0$, $T1$, $D0$ or $D1$. In working out the behaviour of the machine it is convenient first of all to make up a sequence of random digits for use when the U cases occur. Underneath these we may write the sequence of situations, and have other rows for the corresponding entries from the character, and for the actions of the teacher. The character and the values stored in the memory units may be kept on another sheet. The T entries may be made in pencil and the D entries in ink. A bit of the behaviour of the machine is given below:

Random sequence	0 0 1 1 1 0 0 1 0 0 1 1 0 1 1 0 0 0
Situations	3 1 3 1 3 1 3 1 2 4 4 4 3 2 . .
Alternative given	$\quad U\ T\ T\ T\ T\ U\ U\ S\ S\ S\ U\ T$
by	$\quad 0\ 0\ 0\ 0\ 0\quad\ 1\ 1\ 1\quad 0$
Visible action	$B\ A\ B\ A\ B\ A\ B\ A\ B\ A\ A\ A\ B\ B$
Rew., and Pun.	$\qquad\qquad\qquad\quad P$
Changes in $S1$	1 $\qquad\qquad\qquad\qquad$ 0

It will be noticed that the machine very soon got into a repetitive cycle. This became externally visible through the repetitive $BABAB...$ By means of a pain stimulus this cycle was broken.

It is probably possible to organize these P-type machines into universal machines, but it is not easy because of the form of memory available. It would be necessary to organize the randomly distributed 'memory units' to provide a systematic form of memory, and this would not be easy. If, however, we supply the P-type machine with a systematic external memory, this organizing becomes quite feasible. Such a memory could be provided in the form of a tape, and the externally visible operations could include movement to right and left along the tape, and altering the symbol on the tape to 0 or to 1. The sense lines could include one from the symbol on the tape. Alternatively, if the memory were to be finite, e.g. not more than 2^{32} binary digits, we could use a dialling system. (Dialling systems can also be used with an infinite memory, but this is not of much practical interest.) I have succeeded in organizing such a (paper) machine into a universal machine.

The details of the machine involved were as follows: There was a circular memory consisting of 64 squares of which at any moment one was in the machine ('scanned') and motion to right or left were among the 'visible actions'. Changing the symbol on the square was another 'visible action', and the symbol was connected to one of the sense lines $S1$. The even-numbered squares also had another function, they controlled the dialling of information to or from the main memory. This main memory consisted of 2^{32} binary digits. At any moment one of these digits was connected to the sense line $S2$. The digit of the main memory concerned was that indicated by the 32 even positioned digits of the circular memory. Another two of the 'visible actions' were printing 0 or 1 in this square of the main memory. There were also three ordinary memory units and three sense units $S3$, $S4$, $S5$. Also six other externally visible actions A, B, C, D, E, F.

This P-type machine with external memory has, it must be admitted, considerably more 'organization' than say the A-type unorganized machine. Nevertheless, the fact that it can be organized into a universal machine still remains interesting.

The actual technique by which the 'organizing' of the P-type machine was carried through is perhaps a little disappointing. It is not sufficiently analogous to the kind of process by which a child would really be taught. The process actually adopted was first to let the machine run for a long time with continuous application of pain, and with various changes of the sense data $S3$, $S4$, $S5$. Observation of the sequence of externally visible actions for some thousands of moments made it possible to set up a scheme for identifying the situations, i.e. by which one could at any moment find out what the situation was, except that the situations as a whole had been renamed. A similar investigation, with less use of punishment enables one to find the situations which are affected by the sense lines; the data about the situations involving the memory units can also be found but with more difficulty. At this stage the character has been reconstructed. There are no occurrences of $T0$, $T1$, $D0$, $D1$. The next stage is to think up some way of replacing the D's of the character by $D0$, $D1$ in such a way as to give the desired modification. This will normally be possible with the suggested number of situations (1000), memory units etc. The final stage is the conversion of the character into the chosen one. This may be done simply by allowing the machine to wander at random through a sequence of situations, and applying pain stimuli when the wrong choice is made, pleasure stimuli when the right one is made. It is best also to apply pain stimuli when irrelevant choices are made. This is to prevent getting isolated in a ring of irrelevant situations. The machine is now 'ready for use'.

The form of universal machine actually produced in this process was as follows: Each instruction consisted of 128 digits, which we may regard as forming four sets of 3, each of which describes one place in the main memory. These places may be called P, Q, R, S. The meaning of the instruction is that if p is the digit at P and q that at Q then $1 - pq$ is to be transferred to position R and that the next instruction will be found in the 128 digits beginning at S. This gives a U.P.C.M., though with rather less facilities than are available say on the ACE.

I feel that more should be done on these lines. I would like to investigate other types of unorganized machine, and also to try out organizing methods that would be more nearly analogous to our 'methods of education'. I made a start on the latter but found the work altogether too laborious at present. When some electronic machines are in actual operation, I hope that they will make this more feasible. It should be easy to make a model of any particular machine that one wishes to work on within such a U.P.C.M. instead of having to work with a paper machine as at present. If also one decided on quite definite 'teaching policies' these could also be programmed into the machine. One would then allow the whole system to run for an appreciable period, and then break in as a kind of 'inspector of schools' and see what progress had been made. One might also be able to make some progress with unorganized machines more like the A and B types. The work involved with these is altogether too great for pure paper-machine work.

One particular kind of phenomenon I had been hoping to find in connection with the P-type machines. This was the incorporation of old routines into new. One might have 'taught' (i.e. modified or organized) a machine to add (say). Later one might teach it to multiply by small numbers by repeated addition and so arrange matters that the same set of situations which formed the addition routine, as originally taught, was also used in the additions involved in the multiplication. Although I was able to obtain a fairly detailed picture of how this might happen I was not able to do experiments on a sufficient scale for such phenomena to be seen as part of a larger context.

I also hoped to find something rather similar to the 'irregular verbs' which add variety to language. We seem to be quite content that things should not obey too mathematically regular rules. By long experience we can pick up and apply the most complicated rules without being able to enunciate them at all. I rather suspect that a P-type machine without the systematic memory would behave in a rather similar manner because of the randomly distributed memory units. Clearly this could only be verified by very painstaking work; by the very nature of the problem 'mass production' methods like built-in teaching procedures could not help.

12. DISCIPLINE AND INITIATIVE

If the untrained infant's mind is to become an intelligent one, it must acquire both discipline and initiative. So far we have been considering only discipline. To convert a brain or machine into a universal machine is the extremest form of discipline. Without something of this kind one cannot set up proper communication. But discipline is certainly not enough in itself to produce intelligence. That which is required in addition we call initiative. This statement will have to serve as a definition. Our task is to discover the nature of this residue as it occurs in man, and to try and copy it in machines.

Two possible methods of setting about this present themselves. On the one hand we have fully disciplined machines immediately available, or in a matter of months or years, in the form of various U.P.C.M's. We might try to graft some initiative onto these. This would probably take the form of programming the machine to do every kind of job that could be done, as a matter of principle, whether it were economical to do it by machine or not. Bit by bit one would be able to allow the machine to make more and more 'choices' or 'decisions'. One would eventually find it possible to programme it so as to make its behaviour be the logical result of a comparatively small number of general principles. When these became sufficiently general, interference would no longer be necessary, and the machine would have 'grown up'. This may be called the 'direct method'.

The other method is to start with an unorganized machine and to try to bring both discipline and initiative into it at once, i.e.

instead of trying to organize the machine to become a universal machine, to organize it for initiative as well. Both methods should I think be attempted.

INTELLECTUAL, GENETICAL AND CULTURAL SEARCHES

A very typical sort of problem requiring some sort of initiative consists of those of the form 'Find a number n such that   '. This form covers a very great variety of problems. For instance, problems of the form 'See if you can find a way of calculating the function ... which will enable us to obtain the value for arguments ... to accuracy ... within a time ... using the U.P.C.M. ...' are reducible to this form, for the problem is clearly equivalent to that of finding a programme to put on the machine in question, and it is easy to put the programmes into correspondence with the positive integers in such a way that given either the number or the programme the other can easily be found. We should not go far wrong for the time being if we assumed that all problems were reducible to this form. It will be time to think again when something turns up which is obviously not of this form.

The crudest way of dealing with such a problem is to take the integers in order and to test each one to see whether it has the required property, and to go on until one is found which has it. Such a method will only be successful in the simplest cases. For instance, in the case of problems of the kind mentioned above, where one is really searching for a programme, the number required will be somewhere between 2^{1000} and $2^{1000000}$. For practical work, therefore, some more expeditious method is necessary. In a number of cases the following method would be successful. Starting with a U.P.C.M. we first put a programme into it which corresponds to building in a logical system (like Russell's Principia Mathematica). This would not determine the behaviour of the machine completely: at various stages more than one choice as to the next step would be possible. We might, however, arrange to take all possible arrangements of choices in order, and go on until the machine proved a theorem, which, by its form, could be verified to give a solution of the problem. This may be seen to be a conversion of the original problem into another of the same form. Instead of searching through values of the original variable alone searches through values of something else. In practice, when solving problems of the above kind one will probably apply some very complex 'transformation' of the original problem, involving searching through various variables, some more analogous to the original one, some more like a 'search through all proofs'. Further research into intelligence of machinery will probably be very greatly concerned with 'searches' of this kind. We may perhaps call such searches 'intellectual searches'. They might very briefly be defined as 'searches carried out by brains for combinations with particular properties'.

It may be of interest to mention two other kinds of search in this connection. There is the genetical or evolutionary search by which a combination of genes is looked for, the criterion being survival value. The remarkable success of this search confirms to some extent the idea that intellectual activity consists mainly of various lines of search.

The remaining form of search is what I should like to call the 'cultural search'. As I have mentioned, the isolated man does not develop any intellectual power. It is necessary for him to be immersed in an environment of other men, whose techniques he absorbs during the first 20 years of his life. He may then perhaps do a little research of his own and make a very few discoveries which are passed on to other men. From this point of view the search for new techniques must be regarded as carried out by the human community as a whole, rather than by individuals.

13. INTELLIGENCE AS AN EMOTIONAL CONCEPT

The extent to which we regard something as behaving in an intelligent manner is determined as much by our own state of mind and training as by the properties of the object under consideration. If we are able to explain and predict its behaviour or if there seems to be little underlying plan, we have little temptation to imagine intelligence. With the same object therefore it is possible that one man would consider it as intelligent and another would not; the second man would have found out the rules of its behaviour.

It is possible to do a little experiment on these lines, even at the present stage of knowledge. It is not difficult to devise a paper machine which will play a not very bad game of chess. Now get three men as subjects for the experiment A,B,C. A and C are to be rather poor chess players, B is the operator who works the paper machine. (In order that he should be able to work it fairly fast it is advisable that he be both mathematician and chess player.) Two rooms are used with some arrangement for communicating moves, and a game is played between C and either A or the paper machine. C may find it quite difficult to tell which he is playing. (This is a rather idealized form of an experiment I have actually done.)

SUMMARY

The possible ways in which machinery might be made to show intelligent behaviour are discussed. The analogy with the human brain is used as a guiding principle. It is pointed out that the potentialities of the human intelligence can only be realized if suitable education is provided. The investigation mainly centres round an analogous teaching process applied to machines. The idea of an unorganized machine is defined, and it is suggested that the infant human cortex is of this nature. Simple examples of such machines are given, and their education by means of rewards and punishments is discussed. In one case the education process is carried through until the organization is similar to that of an ACE.

REFERENCES

[1] Gödel, K. 'Über formal unentschiedbare Sätze der Principia Mathematica und verwandter Systeme', *Mh. Math. Phys.* *38* (1931), 173-189

[2] Church, Alonzo, 'An unsolvable problem of elementary number theory', *Am. J. Math.*, *58* (1936), 345-363

[3] Turing, A. M., 'On computable numbers with an application to the Entscheidungs-problem', *Proc. Lond. math. Soc.* (2), *42* (1937), 230-265

3

INTELLIGENT BEHAVIOUR IN PROBLEM-SOLVING MACHINES

Ever-present in any discussion of the scope and potential of complex analytical machines, are questions relating to the 'thought processes' of these machines, and their relationship to supposedly similar processes in man. That man should have some momentary difficulty in understanding how machines could exhibit problem-solving behaviour is not surprising—psychologists and physiologists experimented for decades with animals in an effort to see whether they were capable of systematic problem-solving, and took some time before agreeing that they were. In the following paper, Gelernter and Rochester set out to discuss (and ultimately to construct) a machine which will attempt to solve problems in an 'intelligent' fashion, and deal with one specific type of problem—seeking new proofs of Euclidean plane geometry. The machine they describe uses the heuristic approach, not tackling the task by the routine exploration of every conceivable possibility (which is, of course, one way, if an inefficient one, of solving the most difficult problems). Their final speculations, made 10 years ago, make interesting reading. Beyond their 'Theorem Machine', they believe, lies a 'Theory Machine' which should with suitable elaboration turn out to be very good at both basic and imaginative scientific research. The paper also serves to remind one that the question of the nature of 'learning machines' is still open and topical.

Further Reading

Minsky, M. L. (1959). 'Some methods of artificial intelligence and heuristic programming'. *N.P.L. Symposium on the Mechanisation of Thought Processes*, H.M.S.O., London. 5-27.

Tolman, E. C. (1939). 'Predictions of vicarious trial and error by means of the schematic sowbug'. *Psychol. Rev.* **46**. 318-36.

Reproduced from *I.B.M. Journal* 2 (1958) 335-345 by courtesy of the Editor

3

INTELLIGENT BEHAVIOUR IN
PROBLEM-SOLVING MACHINES

H. L. Gelernter and N. Rochester

Abstract: As one step in the study of intelligent behaviour in machines, the authors consider the particular case of a machine that can prove theorems in elementary Euclidean plane geometry. The device uses no advanced decision algorithm, but relies rather on rudimentary mathematics and 'ingenuity' in the manner, for example, of a clever high-school student.
This paper discusses heuristic methods and learning machines and introduces the concept of a theory machine as an extension of a theorem-proving machine.

INTRODUCTION

Modern machines execute giant tasks in arithmetic and carry out clerical operations that are far beyond human capacity, but we have not yet learned to apply computers to problems that require more than a barest minimum of ingenuity or resourcefulness. This paper reports some early results in an approach to the problem of learning how to use machines in these currently unmanageable areas. The goal of this research is the design of a machine whose behaviour exhibits more of the characteristics of human intelligence.

We shall concern ourselves in particular with a single representative problem, one which contains in relatively pure form the difficulties we must understand and overcome in order to attain our stated goal. The special case we have chosen is the proof of theorems in Euclidean plane geometry in the manner, let us say, of a high-school sophomore. It must be emphasized that although plane geometry will yield to a decision algorithm, the proofs offered by the machine will not be of this nature. The methods to be developed will be no less valid for problem solving in systems where no such decision algorithm exists.

If the application of a decision algorithm is rejected as uninteresting (in the case of plane geometry) or impossible (for most problems of interest), there remain two alternative approaches to the proof of theorems in formal systems. The first consists in exhaustively developing the proof from the axioms and hypotheses of the system by systematically applying the rules of transformation until the required proof has been produced (the so-called 'British Museum algorithm' of Newell and Simon[1]). There is ample evidence that this procedure would require an impossibly large number of steps for all but the most trivial theorems of the most trivial formal systems. The remaining alternative is to have the machine rely upon heuristic methods, as people usually do under similar circumstances.

Problems for which people use heuristic methods seem to have the following characteristic. The work begins routinely, and then suddenly the person experiences a flash of understanding. This is followed by the writing down and checking of the solution. Apparently the person first used heuristic methods to look for a solution. To each suggestion turned up by the heuristic methods he applies some sort of a test. The flash of understanding comes when a suggestion gets a high score on the test. The clerical task that follows is the transformation from *suggestion space** to *problem space*. The transformation is possible, of course, only if a valid solution has been indicated. The geometry machine will behave in this way.

Instead of geometry we might have chosen a certain class of probability problems, proofs of theorems in projective geometry, proofs of trigonometric identities, proofs in part of number theory, or the evaluation of indefinite integrals. There were compelling reasons, however, for choosing plane geometry, the most important being the readily understood 'suggestion space' offered by the diagram (the semantic interpretation of the formal system), and the ease of transforming 'proof indications' into problem space. An important secondary reason was the fact that everyone who would be interested in our results has studied Euclid, so the results can be communicated more efficiently.

It should be noted here that the geometry project is a consequence of the Dartmouth Summer Research Project on Artificial Intelligence, standing on a foundation laid by the members of study†, and evolving from the pioneering work of Newell and Simon in heuristic programming[2].

* Newell and Simon have used the term *planning space*.
† Particularly J. McCarthy, M. L. Minsky and one of the authors (N.R.)

Not all problems whose solutions seem to be accompanied by a 'flash of understanding' are elementary enough to lie within the scope of the methods described in this paper. Many have difficulties of a more profound nature. It will be possible to say a little more about this later, but a secure understanding of the nature of these harder problems will come only after more research has been done.

The explanation of the precise meaning of the term *heuristic method* is an important part of this paper. For the moment, however, we shall consider that a heuristic method (or a *heuristic*, to use the noun form) is a procedure that may lead us by a short cut to the goal we seek or it may lead us down a blind alley. It is impossible to predict the end result until the heuristic has been applied and the results checked by some formal process of reasoning. If a method does not have the characteristic that it may lead us astray, we would not call it a heuristic, but rather an algorithm*. The reason for using heuristics instead of algorithms is that they may lead us more quickly to our goal and they allow us to venture by machine into areas where there are no algorithms†.

Finally, since people seem to use heuristic reasoning in nearly every intelligent act, it is reasonable to ask why some task more familiar and natural for people was not chosen as representative of the class rather than plane geometry. Several alternatives to geometry were, in fact, considered and rejected for failing to satisfy one or more of the following requirements:

1. The task must include a kind of reasoning that we are not yet able to get our machines to do but which we think we can learn to manage.

2. It must not contain harder kinds of reasoning that are too far beyond our understanding.

3. It must not be too much cluttered with irrelevant work.

---

* A decision procedure applied under the constraint of a time limit behaves as if it were a heuristic.

† There are classes of problems, for example proofs of theorems in number theory, where it can be shown that no decision procedure can be devised. Heuristic procedures should enable machines to solve problems that are members of such classes. It should be evident that no set of heuristics together with the programme to employ them can guarantee that a machine will solve every member of such a class. All a machine can do is to probe around and perhaps come up with an answer. This, of course, is all that people can do. It should be evident, too, that a programme utilizing heuristics can well be an algorithm that is guaranteed to solve any member of some class of problems. Such a class must, of course, be amenable to a decision procedure. The contribution of an individual heuristic here is that it may lead to a short cut. The geometry theorem machine will probably be an algorithm of this type.

Most human acts fail to meet requirement (2). We have a long way to go before our machines can play Turing's 'Imitation Game' and win[3].

GEOMETRY

A standard dictionary defines geometry as 'the theory of space and of figures in space', and indeed, most people would offer a similar definition. To the mathematician, however, geometry represents a formal mathematical system within which proofs are possible, and which can be related to real space if this seems interesting for the purpose at hand, but which can alternatively be related to concepts having no physical reality or significance. The machine considers geometry primarily as a formal system but uses the interpretation in terms of figures in space for heuristic purposes.

A formal system such as geometry comprises:

1. Primitive symbols
2. Rules of formation
3. Well-formed formulae
4. Axioms
5. Rules of inference
6. Theorems.

The set of primitive symbols (or alphabet) for geometry is those characters which are interpreted as the names of points together with those interpreted as specifying relations between discrete sets of points, or between a given set and the universe of points (e.g., =, $\parallel$, A, B, Δ). In order to make proofs in geometry it is not necessary, for example, to think of a line as something long, thin, and straight. It is sufficient to be able to recognize the symbol *line*.

The rules of formation specify how to assemble the primitive symbols into well-formed formulae (statements) which may be valid or invalid within the formal system. For example, 'Two sides of every triangle are parallel' is a well-formed formula (although not valid), whereas 'Two exists of obtuse every one point' is not a well-formed formula. We can ask the machine whether the first is true (interpreting formal validity as truth), but the second is gibberish because it does not obey the rules of formation. These rules are, in a sense, the grammar of a language whose vocabulary comprises the alphabet of primitive symbols.

The axioms are a set of well-formed formulae, such as 'Through every pair of points there can be drawn one and only one straight line,' which are selected to serve as a foundation on which to build. They are regarded as being true by definition, if you like.

The rules of inference are the means by which the validity of one well-formed formula can be derived from others that are already established. The new formula is said to be inferred immediately from the given one or set by the specified rule of inference.

A proof is a succession of well-formed formulae in which each formula (or line of proof) either follows by one of the rules of inference from the preceding formulae, or is an axiom or previously established theorem. A theorem is the last line in a proof.

To recapitulate, a problem presented to our machine is a statement in a formal logistic system, and the solution to that problem will be a sequence of statements, each of which is a string of symbols in the alphabet of that system. The last statement of the sequence will be the problem itself, the first will always be an axiom or previously established theorem of the system*. Every other formula will be immediately inferable from some set preceding it or will itself be an axiom or previously established theorem.

This simple and elegant description of geometry is essentially the one given the high-school sophomore. It will shortly be seen that this view is too naïve to describe what really happens, but for the moment it will be expedient to continue the exposition as if it were true, because the idealization has a significance of its own. A number of things should be pointed out about this ideal view of geometry. First, there is a difference between finding a proof and checking it. To check a proof one merely follows some simple rules that are set down very precisely. To discover a proof, on the other hand, requires ingenuity and imagination. One must use good intuitive judgment in selecting which of many possible alternatives is a step in the right direction. The high school sophomore does not have a complete set of explicit rules to guide him in finding a proof.

Since checking a proof is a clerical procedure there is no reason why a machine cannot easily do it. A well-formed formula (i.e., axiom, line of a proof, or theorem) would be a string of data words in memory, and a rule of formation or of inference would be a subprogramme. There is nothing really new or difficult

---

*In the case of a theorem contingent upon a set of hypotheses, the proof is developed in an extended system in which the hypotheses are appended to the original set of axioms. The transformation of this categorical proof to the desired hypothetical one is trivial.

about this, and many programmes have been written to make machines do jobs as difficult. The artificial geometer discussed here will have a subprogramme which is an algorithm for checking a proof.

The process of discovering a proof is another matter, and the question of how to get a machine to do it is the subject of this paper. The student or the machine can be given some useful hints but must be provided with a warning that these hints may be misleading. For example, it can be said that if the proposition to be proved involves parallel lines and equality of angles, there is a good chance that it will help to try the theorem: 'If two parallel lines are intersected by a third line, the opposite interior angles are equal'. This advice is a heuristic that can be given to the machine or student. It will lead to a proof in a good many cases but will as often lead nowhere at all.

Thus far, there has been no mention of drawing figures. It is, of course, quite possible to discover a proof in a formal system without interpreting that system, and in the case of geometry, except for the need to discover proofs efficiently, or for applying theorems to practical problems, one need never make a drawing. The creative mathematician, however, generally finds his most valuable insights into a problem by considering a model of the formal system in which the problem is couched. In the case of plane geometry, the model is a diagram, a semantic interpretation of the formal system in which, to quote Euclid, the symbol *point* stands for 'that which has no parts,' a *line* is 'breadthless length,' and so on. The model is so useful an aid for discovering proofs in geometry that few people would attempt a proof without first drawing a diagram, if not physically, then in view of the mind's eye. If a calculated effort is made to avoid spurious coincidences, then one is usually safe in generalizing any statement in the formal system that correctly describes the diagram, with the notable exception of those statements concerning inequalities.

We cannot emphasize too strongly the following point. To serve as a heuristic device in problem solving, the model need not lie in rigorous one-to-one correspondence with the abstract system. *It is only necessary that they correspond in a sufficient number of ways to be useful.* The success of the model in designating correct solutions to problems in that system (solu-'tions that will be checked within the framework of the abstract system) is the only criterion one need apply in judging the suitability of a given model*. If the model is indeed a semantic interpretation of a formal logistic system, then it is most desirable that the interpretation satisfy every axiom of the formal system. But should the interpretation be valid, too, for some richer or poorer formal system, its heuristic value might be impaired, but by no means eliminated.

HEURISTIC METHOD

The proof of theorems in Euclidean plane geometry in the sense described above requires the extensive use of heuristic methods, and it is these methods rather than geometry that are of primary interest to us. The role of geometry is to provide a problem of the right difficulty to permit a thorough development and understanding of the class of heuristics involved.

The steps in a typical application of a heuristic method to theorem proving are the following:

1. Calculate the character† of the theorem.
2. Using the theorem character, calculate the applicable methods and estimate the merit of each.
3. Select the most appropriate method.
4. Try it.
5. In case of failure, cross off this method and return to step (3).
6. In case of success, print the proof and stop.

The character of a theorem (or of any problem) is in essence the machine's description of the theorem (or the problem). In its simplest form, the character may be represented by a vector, each element of which describes a given property of either the syntactic statement of the theorem or its semantic representation. The vector designating the applicable methods and estimated merit of each is a vector function of the character. The figures of merit are, of course, only guesses based initially on the judgment of the programmer, and subsequently modified by the machine in the light of its experience.

---

* Newell and Simon, in private communication with the author, have described an abstract model for a propositional calculus which is not a semantic interpretation, but which in fact, is another formal system in which it is trivially easy to prove the transformed theorems. Since this is a true heuristic, it is not always possible to transform back to the problem space.

† The term *character* was introduced by Minsky (M. L. Minsky, 'Heuristic Aspects of the Artificial Intelligence Problem', Lincoln Laboratory Report 34-35, December 1956) and is to be understood in its dictionary sense. The particular machine representation of a theorem character selected by the authors differs somewhat from that of Minsky.

Defining the term *characteristic* as a given element of the character vector, the following might be introduced as syntactic characteristics of a theorem:

$C_i = 1$ if the hypotheses contain the symbol $\|$, 0 otherwise;

$C_j = 1$ if the consequents of the theorem contain the symbol $\|$, 0 otherwise;

$C_k = 1$ if there exists a permutation of the names of points in the hypotheses that leaves the set of hypotheses unchanged, 0 otherwise; and so on.

Examples of semantic characteristics are the following:

$C_l = n$, where n is the number of axes of symmetry in the diagram;

$C_m = 1$ if two angles of segments are to be proved equal, and they are corresponding elements of congruent triangles, 0 otherwise; and so on.

The rules formalized into the vector function that transforms the character of a problem into a sequence of designated methods of approach and the estimated merit of each will in general fall into two categories. The first will contain those heuristics which operate on the syntactic characteristics of the problem. The second will, in the general case of a problem-solving machine, comprise those rules which operate on the characteristics of the model. For the artificial geometer, these are the semantic characteristics described above.

The problem of strategy and tactics in choosing methods is most important. One obvious strategy mentioned earlier is to explore all alternatives systematically. This is known to be inadequate for many problems and is considered by the authors to be uninteresting, and probably useless, for geometry. The strategy and tactics used by Newell and Simon in their achievement in theorem proving by machines are not adequate for this harder problem on present-day machines. Their proofs were, at most, three or four steps long, and machine time required is probably an exponential function of the number of steps. Clearly the ten-step proofs of geometry will require much more selective heuristics than those adequate for propositional calculus.

The authors have at present a system of strategy and tactics. It does not seem useful to report it in detail at this time because machine experience will probably induce major revisions and improvements. It is clear, however, that the skill with which the machine selects and manipulates methods will distinguish a good machine from a poor one. Since it is impossible to predict the detailed behaviour of so complex an information-processing system as the artificial geometer, it is necessary to write the programme and run the simulation before conclusions can be reached with confidence.

The speed with which a difficult problem can be solved is an essential factor in determining the usefulness of an intelligent machine. This speed cannot be achieved by little steps like inventing faster components. On the scale considered here, a factor of ten is a minor change in speed. Suppose, for example, that a given proof requires ten steps. If for each step, the machine must explore three alternatives, there will be about 20,000 things to consider. A slightly less intelligent machine that must explore six alternatives will have to consider 20,000,000 things. For problems having longer solutions, selectiveness becomes more important exponentially.

SYNTACTIC SYMMETRY

The formal system of plane geometry will be a difficult one for the machine to manipulate. Not only are the alphabet and axiom set both large, but geometry must be formalized in the lower functional calculus, at the very least. The difficulty is compounded, too, by the fact that the predicates of plane geometry exhibit a high degree of symmetry, and a given statement in the system will in general admit of a multiplicity of completely equivalent forms.

These symmetries are at times a painful thing to contend with; they make it necessary that a theorem be considered in every one of its equivalent forms in any attempt to establish a deduction by means of substitution. On the other hand, they are the basis of a powerful new rule, completely syntactic in nature, that simplifies immensely the search for a proof of a theorem displaying these symmetries. The rule will prevent the machine from searching in a circle for useful intermediate steps, or subgoals, to bridge the gap between antecedent and consequent of the theorem to be proved. In effect, it removes from consideration those subgoals which are formally equivalent to some subgoal already incorporated into the structure of the search for a proof.

We shall introduce the rule by an example. Let us consider the following theorem: 'The diagonals of a parellelogram bisect one another' (*Figure I*). To solve the problem, the machine must establish the formulae $AE = EC$ and $BE = ED$. Now it would be most useful if the artificial geometer could recognize, as people usually do, that the proof for the second formula is essentially the same as that for the first, and therefore only one of the two need be established. But it is even more important that the machine does not fall into the class of trap illustrated by the following redundant search process. The method chosen is that of congruent triangles, and in order to establish the formula $\Delta I \cong \Delta II$, from which the theorem may be immediately inferred, the machine sets at some later stage the subgoal $\Delta III \cong \Delta IV$. The geometer will, in fact, satisfy our requirements on both these points. The mechanism whereby this is accomplished is an embodiment of the theorem and rule specified below.

Consider first the following definition: Let π be a permutation on the names of the syntactic variables in a theorem. Then π is a

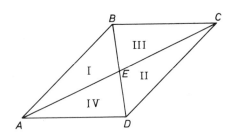

Figure 1

syntactic symmetry of the theorem if its operation on the set of hypotheses leaves the set unchanged except for a possible transformation into an equivalent form with respect to the symmetries of the predicates (i.e., $\pi\{H\} \equiv \{H\}$ is valid). We can now state the required theorem thus:

'If Γ is a well-formed formula provable from the set of hypotheses $\{H\}$, and π is a syntactic symmetry of the set $\{H\}$, then $\pi\Gamma$ is a well-formed formula provable from the same set $\{H\}$. The formula $\pi\Gamma$ will be called a *syntactic conjugate* of Γ.'

The proof of the theorem is quite trivial, and follows from the fact that the syntactic variables in a theorem may be renamed without destroying the validity of the theorem. Thus, if

$\{H\} \supset \Gamma$ is valid, then
$\pi\{H\} \supset \pi\Gamma$ follows by the rule of substitution.
Since $\pi\{H\} \equiv \{H\}$,
$\{H\} \supset \pi\Gamma$ is valid.

The theorem itself grants the machine the same power the human mathematician has at his disposal when he recognizes the equivalence of two different statements with respect to a given formal system, for now it may establish the syntactic conjugate of any valid formula Γ by merely asserting 'similarly, $\pi\Gamma$'. The rule of syntactic symmetry follows from the theorem. It is used by the machine to construct, given the heuristics and methods at its disposal, the optimum *problem-solving graph*, and a description of such a graph follows. (See *Figure 2.*)

Let G_0 be the formal statement to be established by the proof.

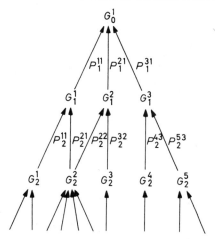

Figure 2. Problem solving graph

The nodes of $G_i{}^\alpha$ represent subgoals of order i, with α numbering the subgoals of a given order. $P_i\alpha\beta$ is a transformation on G into $G_{i-1}\beta$.

It will be called the problem goal. If G_i is a formal statement with the property that G_{i-1} may be immediately inferred from G_i, then G_i is said to be a *subgoal* of order i for the problem. All G_j such that $j<i$ are *higher subgoals* than G_i, where G_0 is considered to be a subgoal of order zero. The *problem-solving graph* has as nodes the G_i, with each G_i joined to at least one G_{i-1} by a directed link. Each link represents a given transformation from G_i to G_{i-1}. The problem is solved when any G can be immediately inferred from the hypotheses and axioms*.

We can now specify the rule of syntactic symmetry thus: G is not a suitable subgoal to add to the problem-solving graph if it is the syntactic conjugate of any G_j for $i \geqslant j$, for any proof sequence leading to G_i is identical with a conjugate sequence leading to G_i with the variables renamed, and any mechanism leading to a

---

* The completed proof will use a deduction metatheorem to get
$$\vdash \{H\} \supset G_0 \text{ from } \{H\} \vdash G_0$$

proof of G_i would as well prove G_j. If $i = j$, the two subgoals are in effect redundant, and if $i > j$, the sequence leading to G_i leads to G_j when conjugated, and all the steps G_k, $i \geqslant k \geqslant j$ can be eliminated.

In the light of the above, we may now re-examine our introductory problem (*Figure 1*). The machine must establish the following two goals:

$$G_0{}^1: \ AE = EC$$
$$G_0{}^2: \ BE = ED$$

By the theorem of syntactic symmetry, the machine will eliminate $G_0{}^2$ from the graph, since $G_0{}^2 = \pi G_0{}^1$, where π is the transformation A *into* B, B *into* C, C *into* D, *and* D *into* A, and after proving $G_0{}^1$, will assert 'similarly, $G_0{}^2$'. Then, if at some point in the proof, $\triangle ABE \cong \triangle CED$ is a subgoal, it will eliminate the statement $\triangle BCE \cong \triangle DEA$ as a possible subgoal; if $AB = CD$ is a subgoal, $BC = DA$ will be removed from consideration. Clearly every directed path through the problem-solving graph from hypotheses to goal will be unique under the π-transformation, and will be the shortest one in that it will contain no redundant subgraphs (no two nodes will be linkable by a π-transformation).

Syntactic rules such as the above will be essential to the success of the plane geometry machine. But while they ease the labour of the geometer considerably as it threads a path from problem to solution, they are, except in the simplest cases, powerless to indicate which path, among the very many possible, does indeed lead to a solution, and which wander off into infinity, regressing farther from the goal with each step. The geometer will need more information about most problems before it can even begin to seek a solution. It will find this information as the mathematician does, in the diagram.

SEMANTIC HEURISTIC

Semantic heuristic is concerned with the body of pertinent and probably true statements that can be obtained by observing the diagram. For example, one of the first such rules to be applied by the geometer in a particular case will be the following:

If the diagram consists of a 'bare' simple polygon, a construction will probably be required.

A rule to indicate which construction to make might be:

If the figure has one axis of symmetry, and it is not drawn, then draw it.

A most useful rule will be:

If the theorem asks that two line segments or angles be proved equal, determine by measuring whether these are corresponding parts of apparently congruent triangles. If so, attempt to prove the congruence. If necessary, draw lines connecting existing points in the diagram in order to create the congruent triangles.

Another frequently used heuristic will be:

If two apparently parallel lines are crossed by a transversal, attempt to establish the parallelism by considering the angles.

A more complete understanding and appraisal of the appropriate heuristics will be one of the major consequences of experimentation.

It should be clear that the best set of heuristic rules, the best compromise between conciseness and efficiency, should not be expected to yield the best proof in every case. Indeed in a number of awkward cases the rules will impede, rather than aid, the search for a concise proof. In some cases the machine will make a construction and produce an elaborate proof while missing a simple, elegant one. People, too, do this. But these awkward cases should be the exception, and the heuristic rules appear sufficiently powerful to make an efficient machine.

RIGOUR

Mathematical rigour becomes a significant matter in two different aspects of the artificial geometer. One of these is that machines can provide, in a sense, more rigorous proofs than have hitherto been available. More important than this is the second aspect—that our machine is like a good human mathematician in that it increases its output and improves its communication with other mathematicians by taking chances with rigour.

Axioms and theorems are objects that can be examined and manipulated by people and machines. These present no problem. However, methods of inference are instructions to do something. In the case of machines they are programmes of instructions in machine language. In the case of people they are instructions expressed in a natural language and intended to control human behaviour. Except for undetected blunders in the design of a machine or in the writing of a set of machine instructions, the machine and its instructions are fully understood. And when one of these blunders is detected, it causes merely annoyance and not bewilderment. Therefore when a machine proves a theorem, there is in principle no doubt about what is going on and, except for possible apprehensions about human blunders or undetected machine malfunctions, there is no doubt about rigour.

It is interesting to observe that the most rigorous treatments of the foundations of mathematics seem equivalent to designing a machine and a machine language and henceforth communicating in this language. In one case[4] the mathematician even uses the term *machine*, although his machines could not actually be built because they contain parts with infinite dimensions. Other really

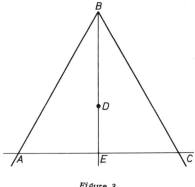

Figure 3

good treatments do not use the word *machine* but are essentially equivalent. It should be clear, then, that the translation of a formal system into a machine programme is reasonable and natural.

The other aspect of rigour is quite different. Most elementary textbooks on geometry fail to prove betweenness relations. In *Figure 3*, the acute angle *ABC* is bisected by the line segment *BD*. Then line segments *BD* and *AC* are extended to infinity, thus becoming lines *BD* and *AC*. Point *E* is defined as the intersection of lines *BD* and *AC*. Now how can it be determined whether point *E* lies between *A* and *C* or to the left of *A* or to the right of *C*?

Ordinarily this decision is made by looking at the figure. In rigorous treatments it is proven formally, but this is a tedious effort. Expediency dictates that the mathematician should neglect the possibility that a semantic heuristic will lead him seriously astray. Since people rarely get into trouble because of honest errors of this kind, traditional geometry excludes proofs of betweenness, and most mathematics appears to lack rigour because many matters are settled by heuristic methods rather than formal proofs. It seems clear that the machine must be able to work this way if it is to become proficient.

The artificial geometer decides questions of betweenness by

measurements on the figures. But whenever it does so, it explicitly records the necessary assumptions for a given proof so as to leave a record of its guesses. There is, of course, a danger that the machine will be proving only a special instance of the theorem presented to it, but this danger can be minimized by having the machine draw alternate diagrams to test the generality of its assumptions when they are necessary.

PROGRAMMING THE GEOMETER

The organization of the programme falls naturally into three parts: a *syntax computer* and a *diagram computer* embedded in an executive routine, the *heuristic computer*. The flow of control is indicated in *Figure 4*. The syntax computer contains the formal system, and its purpose is to establish the proof. The formal system manipulated by the syntax computer is expressed as a Post-Rosenbloom canonical language, and consequently the syntax computer should be useful for a wide range of logistic systems. The heuristic computer can submit any sequence of lines of proof to the syntax computer, which will test them to see that they are correct.

The diagram computer makes constructions and measures them. It does this by means of co-ordinate geometry and floating-point

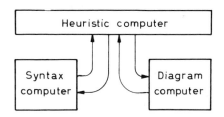

Figure 4. Flow chart of the artificial geometer

calculations. However, it keeps all this secret from the heuristic computer and reports only qualitative information of the type acquired by a mathematician in scanning a well-drawn figure. The behaviour of the heuristic computer and the syntax computer would not be changed if the diagram computer were replaced by a machine that could draw diagrams on paper and observe them.

The heuristic computer does most of the things that have been discussed in this report. It contains the heuristic rules and decides what to do next. The subordinate computers only follow its instructions and answer its questions.

The programme is being written in an information-processing language constructed by appending a large set of special functions to the Fortran compiler for the IBM 704. The language increases manyfold the ease of writing programmes of the nature of the geometer, and will be reported upon in detail in a subsequent paper.

LEARNING IN INTELLIGENT MACHINES

The machine described thus far will exhibit intelligent behaviour but will not improve its technique. Except for the annexation of previously proved theorems to its axiom list, its structure is static. A rigorous sequence of practice problems will not improve its performance at all in solving a given problem unless a usable theorem is among them. Such a machine, incapable of developing its own structure, will always be limited in the class of problems it can solve by the initial intent of its designer. It seems that the problem of designing a machine of general intelligence will be enormously greater, if at all possible, than designing one not so intelligent but with the capacity to learn .

One might attempt to endow an automaton like the geometer with the ability to learn at various levels of sophistication. Indeed, the behaviour of the machine in storing away for future use each theorem it has proved may be interpreted as learning of a rudimentary sort. This might be refined by having the machine become selective in its choice of theorems for permanent storage, rejecting those which (by some well-defined criteria) do not seem to be sufficiently 'interesting' or general to be useful later on. Similarly, instead of 'forgetting' all lemmas it might have established as intermediate steps in the proof of the theorems offered to it, to be rederived when needed, the machine might select the especially interesting ones for its list of established theorems.

The next level of learning is indicated when the machine adjusts, on the basis of its experience, the probability for success it assigns to a given heuristic rule for a theorem with a given character. This is the learning involved when the machine uses results on one problem to improve its guesses about similar problems. As the geometer is given problems of a given class, say problems about parallelograms, its ability to handle them would improve. After it had been given a graded sequence of harder and harder problems, its performance should be much better and it could be said to have learned to prove parallelogram theorems. The highest level to which we aspire for an early model of the geometer will be achieved when it looks over the quality of its predictions and discards as irrelevant some of the criteria that comprise the problem character. The earliest models of the geometer will include only low levels; later models will be more complex.

Beyond these kinds of learning we can see other things. Before we come to them, however, we will probably be working on machines to solve harder problems than those of geometry. There are kinds of learning that are needed only by machines that take their environment more seriously than do theorem-proving machines. But we can hope that a theorem machine might some day be able to observe that a certain sequence of methods was effective in certain circumstances, and consequently streamline the sequence into a single method and in this way devise a new method.

But in still another vein there are possibilities for theorem machines. Instead of providing a machine with a formal system and a sequence of propositions to prove, one could give the machine a formal system and ask it to see what it could find. Here it would at least need criteria for the utility of theorems in proving other theorems and for the elegance of a proof in terms of large achievements in a small number of steps. New kinds of learning would be used here.

Before closing the subject of learning machines, there are some further considerations to deal with. A computer is, after all, merely a finite automaton, and, as such, its behaviour is completely determined by its internal state at the beginning and by subsequent input information This being the case, it can be argued that its response to any set of input signals is in principle predictable and is consequently uninteresting and not worthy of the description 'intelligent'. Another version of this objection is the following. The machine, endowed with heuristics and judgments of its designer, is but a trivial extension of that person, in principle no different from a slide rule in the hands of an engineer.

From a certain irrelevant point of view the objection is justified, but in practice the behaviour of the machine is far from being predictable. That this is indeed the case is well illustrated by the fact that the geometer, its operation simulated 'by hand', has on several occasions produced a proof that was a complete surprise to its programmers. The nature of an intelligent programme is such that unlike a conventional arithmetic computation, in which the branches are few and easily traceable, the

71

number of conditional branches depending on the input are bewilderingly many and highly interdependent, rendering impossible any detailed attempt to trace its behaviour. And of course, once learning is introduced into the programme, it will constantly modify itself in a highly complex way, so that while its behaviour is still in principle determined, one will become increasingly powerless to predict its response in any given case. In a very real sense, the machine's proofs will be no more or less trivial than those offered by the neophyte mathematician who is still under the influence of his professor.

One may view this machine in still another way. At any instant of time, the internal configuration of our machine is some particular state of a finite-state automaton. Then of the infinite number of sequences that one might ask the machine to establish as theorems, some infinite subset of these will be provable by it. At any given time, our machine represents a partial decision method over this infinite set of theorems, and this set will be richer in 'interesting' theorems than a random subset of all theorems. The class of theorems considered 'interesting' will determine the heuristics that control the partial decision method, and in turn, the density of interesting theorems in the set enumerated by the machine will depend on the apt choice of the heuristics. It is important to note that if even the most rudimentary learning behaviour is built into the machine, its initial internal configuration will be different for each new problem presented to it, and consequently, the class of theorems decidable by the machine will be continually changing. And what is any human mathematician but a partial decision machine over some unknown class of theorems?

It is possible to approach the problem of theorem-proving by machine from a rather different direction. E. W. Beth[5] describes a method (semantic tableaux) for systematically constructing a counter-example for a proposed theorem if there is one, or else establishing the fact that none exists. If it can be shown that a counter-example cannot be constructed, an algorithm is given for converting the 'closed' semantic tableau produced into a proof of the theorem in the formal system. But the method of semantic tableaux is essentially an enumeration procedure—in this case, it is the set of individual instances of the theorem that could possibly be counter-examples to the theorem that is being enumerated, and like all such procedures, the bulk of calculation required rapidly outdistances the capacity of conceivable computing machines. In order to make the procedure reasonably efficient, heuristic rules for the control of the enumeration must be introduced, and one is faced with essentially the same problem that concerns the body of this paper. The more or less anthropomorphic approach followed by the authors has the advantage that suitable heuristics are readily suggested by introspection, and the methods developed are more likely to be applicable to the solution of problems in non-formal systems.

THE THEORY MACHINE

At various points in the preceding discussion, a line of reasoning was terminated by the comment that harder problems exist but they are outside the scope of the matter being considered. This large new class of problems, and how a machine can handle them, is the subject of this section. We consider now a machine that takes its environment more seriously.

The subject will be introduced by an example of a more advanced kind of geometry machine, a machine that tries to learn what kind of geometry fits the environment it finds around it. The heuristic computer is provided with an environment by the diagram computer. It looks to the environment for heuristic—for clues about what to do next. However, if it learns that a measurement contradicts something it can prove in the syntax computer, it assumes the measurement is in error. In other words the formal system is sacrosanct.

Now suppose the diagram computer is replaced with another that does its drawings on the surface of a sphere. Suppose further that the priorities in the heuristic computer are readjusted so that it believes the diagram computer rather than the syntax computer when the two are in conflict. Suppose also that it is provided with the means to modify the formal system and additional heuristic to enable it to do so efficiently. It would be arranged so as to try to bring theory (the syntax computer) and experiment (the diagram computer) into harmony, and thereby discover what kind of a world it lives in. This is a theory machine.

There seems to be, in principle, no reason why a theory machine should not be fitted with the means to do experimentation, with a tool room, a stockroom, and an instrument room, and told to work out the theory of something or other. In practice, there is the familiar difficulty of speed and cost. Today it is cheaper and quicker to use people to do research, but perhaps some day machines will do the research and people will merely control the doing of research. This is precisely parallel to the digging of excavations. At one time excavating was manual labour, but now machines do the digging and people merely control the machines. The scientist using a machine to do research would have a role analogous to that of a university professor directing his graduate students.

A further conjecture along this line relates to programming. A person finds it much easier to communicate a complex message to another person than to a machine. Speaking is relaxed and easy, while writing a programme of machine instructions is detailed and exacting. When one person listens to another he often fails to interpret some word correctly for a while, but later other words enable him to understand the earlier word. It seems as if the listener is continually generating hypotheses about what the speaker means and is continually checking these hypotheses and accepting them or rejecting them and casting about for others. In terms of human activity, theorizing is much too pretentious a word for this activity. However, from the point of view of machine design, it may be that only a theory machine will be easy for people to instruct.

The interaction between formal and heuristic procedures in a theory machine is more intricate than in a theorem machine. To determine the consequences of its present hypothesis the theory machine must use the methods of the theorem machine. Because of the different nature of the typical problems it will be solving, the theory machine must lean more heavily on semantic heuristic as a substitute for rigorous deduction. Then when it finds a discrepancy between theory and experiment it must use both rigorous deduction and heuristic procedures to modify its formal system. An interesting feature of such a machine is that the rules for formal deduction used to modify the formal system are actually part of the formal system. This is not an unreasonable situation; it is essentially what happens when the programme for a calculator causes the calculator to modify the programme. But it surely is complicated, and the complication does not end here.

The machine described so far resembles a theoretician with little or no experimental skill. Additional heuristic is required to enable the machine to select a 'clean' experiment that will be an effective test of a theory. Contingencies will arise in the experimentation, and the machine must handle these as subproblems. In other words it must invoke this whole apparatus over again at a lower level.

The theory machine is a device that conjectures about its environment and tests its conjectures. In so doing, it gains an increased understanding of what is going on. It is hoped that not only will the theory machine be able to do research, but will also be easier to communicate with than a present-day automatic calculator.

SUMMARY

In contrast to the present use of automatic calculators which outperform human beings in clerical tasks, the theorem machine is proposed as a device that reasons heuristically. It is therefore able to solve harder problems, and the study of it reveals some things about the nature of problems and of machines. The essential operating principle of this kind of artificial intelligence is that it has a formal part, a syntax computer that can make deductions, and a heuristic part that can make guesses. By using the syntax computer to test the guesses made on a heuristic basis, the machine is able to get results that are beyond the scope of a purely deductive machine.

Heuristic processes can be syntactic, whereby they depend on the language in which the problem is stated, and on the statement in that language, or they can be semantic and depend upon an interpretation or model of the formal system.

The artificial geometer is an example of a theorem machine. Geometry was chosen, not because of any inherent interest, but rather because it provides an example of a problem at the right level of difficulty that needs semantic heuristic in a major way. The geometer is being studied by simulation on the IBM 704 Electronic Data Processing Machine.

An interesting aspect of the geometry taught in high school is that it is not rigorous. Some facts are established by proving them and some by observing the figure (i.e., semantic heuristic). This is a powerful, effective method of reasoning used by people and by the artificial geometer. While it would be possible, and probably easier, to make the artificial geometer perfectly rigorous, it is more significant in the study of artificial intelligence to avoid the strictness of rigour that is a proper part of metamathematics but not efficient in mathematics.

Beyond the theorem machine is the theory machine which, by conjecturing and testing the conjectures, gains an understanding of its environment. Such a machine should be able to do research and should be easier to communicate with.

The largest obstacle to the development of useful theorem and theory machines is the problem of speed. This cannot be cured by faster components alone. The major contribution to speed must come from improved heuristic so that the machine will waste less time in fruitless endeavour. The nature of hard problems insures that the machine must waste some time on wrong hunches, but the waste must be kept within bounds. The machines themselves are expected to make a major contribution to the understanding of artificial intelligence because they learn as they work, and what they learn reveals much.

ACKNOWLEDGMENT

The authors wish to acknowledge the contributions of A. Newell,
J. McCarthy, M. L. Minsky, and H. A. Simon, whose relation to
the project has been indicated in the text, and to C. L. Gerberich,
J. R. Hansen, and R. M. Krause, whose technical and programming contributions are making the project possible. Professor McCarthy, in particular, has been playing a continuous
role as consultant to the authors.

REFERENCES

[1] Newell, A., Shaw,   . and Simon, H. A., 'Empirical Explorations of the
Logic Theory Machine. A Case Study of Heuristic,' *Proc. west.
Computer Conf.,* p. 218 (February, 1957).
[2] Newell, A. and Simon, H. A., *I.R.E. Trans. Inf. Theory,* IT-2, No. 3,
61 (1956).
[3] Turing, A. M. 'Computing Machinery and Intelligence,' *Mind, 59*
(1950) 433.
[4] Turing, A. M. *Proc. Lond. Math. Soc.,* Series 2, 24 (1936) 230-265.
[5] Beth, E. W. 'Semantic Entailment and Formal Derivability,' *Mede-
delingen der Koninklijke Nederlandse Akademie van Wetenshappen,
afd. Letterkunde, Nieuwe Reeks, 18,* No. 13, 309 (1955). *See also:*
Abraham Robinson, 'Proving a Theorem (as done by Man, Logician, or
Machine),' *Transcription of the Proceedings of the 1957 Cornell
Summer Institute of Logic,* Ithaca, New York

4

MACHINES, ROBOTS AND MINDS

In this short essay, written in 1954 Thomson and Slucking deal succinctly with the kinds of problems and questions raised in the minds of students when first faced with the subject of cybernetics—in particular the nature of ultra-intelligent machinery. In a sense this paper presents a nice complement to Turing's paper published in 1948 and included in this present volume (pp. 79-87). Six 'special properties' of living brains are discussed—Purposeful Behaviour, Consciousness, Thinking, Learning, Memory, and Perception,—and related to possible parallel functions in logically predictable (or even existing) machines. The paper is technically out of date, but in fact all its basic arguments are still quite valid. It was written at a time when the sub-science of Bionics (the study of living systems with special reference to the construction of cybernetic artefacts) was in full flood. Because of excessive idealism, and a consequent failure to fulfil its promises, Bionics went into a ten-year slump from which it is now however beginning to emerge again.

Further Reading
Polanyi, M. (1952). 'The hypothesis of cybernetics'. *Br. J. Phil. Sci.* **11**, 312-15.
Lashley, K. S. (1963). 'Some Reflections on Brain and Mind'. *Brain* **86**, 381-402.

Reproduced from *Durham University Journal* 46 (1954) 116-122 by courtesy of the Editor.

MACHINES, ROBOTS AND MINDS

R. Thomson and W. Sluckin

In 1923, McDougall published the first edition of *An Outline of Psychology*. In this work McDougall gives seven 'marks of behaviour' which define the difference between living and inanimate things. Were he living today, McDougall would be obliged to revise his criteria, since purposeful action and learning, as he described these capacities, are now capable of being reproduced mechanically. It might be supposed that McDougall's error was simply verbal, a matter of inadequate definitions. This is not altogether the case: facts, as well as definitions, form the content of the discussions on teleological mechanisms which have taken place within the last 15 years and received the general title 'Cybernetics'.*

In these writings adaptability in living organisms has been analysed in terms of the physical concepts of stable equilibrium, negative feed-back† and, more recently, ultrastability[1]. The general conclusion is that the traditional antithesis between teleological and mechanical causation is resolved since teleological causation can be *explained* fully in mechanical terms. The technical aspects of these discussions will not concern us in this article. Instead we intend to confine it to one aspect of Cybernetics: namely, the conclusions which are reached by cyberneticians when they apply their thoughts to the nature of the human mind. The mathematicians, electronic engineers, neurophysiologists and others who produce this kind of literature are inclined to defend a mechanistic metaphysic of mind as part of their general philosophy of 'scientific materialism'. (There are, of course, some among them who attack this materialism.)

---

* From the Greek $\kappa v \beta \epsilon \rho v \alpha v$ (Latin *gubernare*) 'to control' or 'govern'.
† When part of the output of an electrical device is fed back 'negatively' into the input, decreasing the ultimate output but improving its stability; see Wiener, N. *Cybernetics*, Wiley, New York, 1948.

The discussion centres on the relevance of the facts about electronic artefacts to the materialist philosophy of mind. Our purpose in this article is to suggest that much of the empirical content in Cybernetics is irrelevant to these metaphysical problems, although we suggest that certain analogies are not merely evocative of traditional metaphysical discussions (namely, sections IV, V and VI). We cannot give a thorough appraisal of the subject within the limits of this article: our aim is to select a few of the more generally discussed topics as illustrations.

1. PURPOSEFUL BEHAVIOUR

Ross Ashby and Norbert Wiener are convinced that the functioning of certain electronic machines is describable as 'teleological' in exactly the same sense in which we regard the functioning of living things as 'teleological'. On analysis, the dispute between those who claim that machines exhibit purposes and those who deny this proposition depends upon the meanings which may be given to certain terms, namely: 'purpose', 'goal-seeking', 'deliberate', 'activity', 'movement'. The entire family of teleological expressions turns out, on inspection to be both ambiguous and vague (in the logicians' use of these terms). As the authors have pointed out[2] there are certain uses of 'purpose' which cannot be applied to robots: the only clear analogy is between human dispositional behaviour (the acquisition of skills and the making of intelligent adaptations to changing situations in the environment) and robot simulations of this kind of behaviour. To what extent the analogy is a close one can be determined by careful observation and description. Thereafter the matter is a question of the 'proper use' of words and phrases. The extent of the ambiguity and vagueness which this particular family of terms reveals, makes clear-cut decisions about their 'proper use' almost impossible. (cf. Thomson, *op. cit.* pp. 138-143.)

It is evident from an examination of the literature that the dispute 'Do robots exhibit purposeful behaviour or do they not?' proceeds in three stages. (1) Descriptions are given of what robots do and how they do it. (2) This is followed by an analysis of what we mean when we apply concepts of intelligence to living creatures. (3) Illustrations are provided with a view to persuading us to extend our use of the concepts of intelligence, previously applied only to human beings, to robot performances. The dispute is over what is said in 2 and 3, and here we are *not* engaged in the examination of empirical evidence but in arguments over how words ought to be used. Such arguments are not unimportant. The aesthetic and expressive use of language is not its only use: the primary purpose of the intolerable wrestle with words and meanings is to improve deliberately on our skill in the use of language, to minimize confusion in our thought and to reduce the possibilities of error and misunderstanding. However, much of what is written in Cybernetics is not a clear-cut examination of the way in which certain words function, or may be allowed to function, but a careless mishandling of psychological words which results in parodox.

Of course, there is usually some point in paradox and in the queer arguments which promote paradox: there is some point in saying that 'robots have minds in the same sense in which human beings have minds' or 'human beings are merely complicated robots', just as there is some point in saying 'human beings are immortal spirits' or 'in every heart a Devil sleeps'. It is one thing to allow such expressions to move us emotionally; it is another thing to attempt to interpret them, although how this is best done is no easy matter to decide. The business of attempting the most complete interpretation of such conclusions and of the reasonings which lead to them is traditionally the task of the philosophers. Cybernetics, in discussing the capacity of machines to act purposefully, is merely supplying the philosopher with a new version of an old paradox.

2. CONSCIOUSNESS

Cybernetics revives the old worry over the relation between mental phenomena and physical events in the brain. In a recent article, an appeal is made to the fact of 'consciousness' as a ground for rejecting the 'mechanistic' view of mind encouraged by cybernetics[3]. Again the question may be asked 'What is the relevance of the facts about robots to this particular metaphysical puzzle?' The worry about the nature of 'consciousness' may be tackled in two ways.

(a) Philosophical analysis of such terms as 'consciousness', 'experience' and 'awareness' suggests that such words do not refer to some attribute of a person, which psychologists cannot ever locate, describe, or analyse, however hard they try. It may be argued[4] that sentences containing these words (or any sentences which imply that a human being experiences mental states or events) can be interpreted as saying something about behaviour. Puzzles about 'raw feels' (the essential experience of, for example apprehending a red patch) arise out of the awkwardness of certain concepts and our tendencies to mis-handle them, and not out of our failure to discover the relevant facts[5].

(b) Even if we were to dispute the results of the philosophical analysis of the problem of 'consciousness', this problem need not concern scientific workers or psychologists and neuro-physiologists. It does not matter whether a psychologist asserts or denies that there are mental events as well as physical ones, or whether 'experience' is something over and above responses. Such metaphysical disputation is irrelevant to the job of framing and testing hypotheses. Experimental psychologists seem generally disposed to accept this part of the Behaviourist pro-gramme—the methodological maxim that attempts to investigate, describe and analyse the contents of consciousness lead to a sterile and useless science. If the concept of consciousness is to be brought into psychology at all, and there is no reason why it should not, it must be defined in terms of behaviour, for example 'a differential response to stimuli', a 'behaviour-readiness', or 'a readiness to discriminate'.

The problem of consciousness is to be treated by logical and semantic methods. In so far as Cyberneticians do anything else, they are not making any essential contribution in this field. Facts about the functioning of the brain and central nervous system, facts about the functioning of robots, and the drawing of analogies between these two sets of facts—these considerations are irrelevant to the philosophical problem of showing what 'consciousness' means and how the notion confuses and mis-leads us when we use it in certain contexts.

3. THINKING

It is claimed that robots throw light upon the nature of human thinking. When we consider what they can do and how they do it. we must conclude, it is suggested, that they ought to be des-cribed as 'thinking'. This claim is disputed. The dispute may be shown to be concerned with the legitimate use of the concepts involved in the classification of a new set of facts (what robots can do). The dispute can be clarified only by the careful analysis of the principles governing the use of concepts of mentality. Such problems concern epistemologists and semanticists and need not be the immediate concern of psychologists.

Whatever meanings the concepts involved may have, or acquire in the development of our language, psychologists are bound to employ only those uses of a term which can be defined in terms of behaviour (e.g., in the case of 'thinking', such definitions as 'problem-solving' or 'adaptive response'). If psychologists wish to study thought processes by testing hypotheses, 'thinking' must be given such definition; it is no use employing a sense of 'think' which means, for example, 'How it feels to X when X poses and solves a problem'. Cybernetics does however influence the thinking of psychologists about 'thinking' in a more indirect manner.

A psychological discussion of the nature of conceptual thinking tends to be concerned with the investigation of thinking as a capacity of the organism. This leads to precise definitions of thinking, and might be aided by the creation of theoretical models fitting such definitions. It may be argued that this is what Spearman and the Gestalt psychologists attempted to do in their deliberations about thought. This is what Rignano did around 1920 when he argued that thinking is imaginative experimentation, or trial and error in imagination. Dewey's analysis of reflective thought supports this view. It is suggested that a thought process consists of a series of guesses or hypotheses which are put one after another to a test of the 'Will it work?' type. Its creativeness is associated with the character of the guesses.

Reasoning is, of course, precise and mathematically and logically analysable, and therefore it is in principle imitated by artefacts, such as Jevons' 'logical piano' or the more recent 'logical-truth calculators'. The *constructive* features of thinking consist of guesses which are *more* or *less* wild, and which are, one after another, put to immediate pragmatic test.

MacKay's model of thinking associated with the conceptual framework of Selective Information Theory, develops this approach[6]. Just as a correct response to a stimulus is gradually arrived at by the elimination of unrewarding responses, so also the correct apprehension of universals is reached by the narrowing down of the range of responses. In principle it is possible to build a mechanism to do what is required to fulfil the conditions of a precise and not inadequate definition of the thought process, although technical difficulties and expense involved deter people from applying themselves to this task.

4. LEARNING

Problems associated with learning belong to the very core of psychology. Do modern mechanisms help us to understand learning? We cannot pretend in the present state of our knowledge to be able to answer this question by a direct 'Yes' or 'No'. But it certainly is not inconceivable that a study of 'mechanical learning' may throw some light on learning in living creatures.

Some types of learning, or some kinds of learning situations, have been imitated by machines; others, in theory, can be. Conditioned reflex is relatively simple to imitate[7]. Trial-and-error learning in a simple form was imitated as far back as 1937 and much more elegantly quite recently[8,9].

Howard's mechanical model follows in the solving of a maze by trial and error the procedure of systematic exploration. Random exploration, wasteful as this method would be, could also ultimately lead to the solving of a maze. A rat running a 'closed maze', that is one where the goal and route are not open to inspection, is half-way; its exploration is neither fully systematic nor entirely random.

A mechanical rat learns a maze perfectly during the first or trial run. On subsequent runs it will make no errors. A real rat learns a maze gradually; it has to run a maze a number of times before it succeeds in eliminating all errors. Now, theoretically, though perhaps not easily in practice, a probability device' could be built into a mechanical maze runner. An error made would then not entirely ensure a correct move on a subsequent occasion. It would merely increase by a given amount the probability of a correct move on a subsequent occasion. Fewer and fewer mistakes would be made during each run, until errors would become very unlikely, and correct, straight-through runs assured.

It is a matter of speculation whether a further development of various other sophisticated mechanisms might not help to settle some of the current controversies in learning theory. Might not progress in learning-model construction help to bridge the gaps between the seemingly conflicting explanations of particular learning behaviours? Might it not be possible to devise a mechanism of learning in different ways according to the contingency of the experimental situation, and in this way help towards the understanding of the true nature of the seemingly different kinds of learning?

5. MEMORY

The character of memory traces has been a subject of debate for many years. The problem has neurophysiological and psychological aspects. The precise nature of the minute but enduring changes in the brain associated with the formation of memories remains elusive. The mechanisms of recognition have been subject to much interesting but inconclusive speculation. The most puzzling features of memory concern its psychological concomitants. Cybernetic discussion has brought forward new ideas in this field, but it may be agreed with Tustin that 'one name for the unknown is as good as another'. Nevertheless it may be wondered whether the new ideas might not prove to be of some value.

For many years now Lashley and his co-workers have been finding that it is almost impossible to localize memory traces in particular parts of the brain[10]. Penfield's researches have not seriously disputed this difficulty. No mechanical storage system seems as suitable as a model for memory because a mechanical trace of any conceivable kind requires special localization. On the other hand, some electronic storage systems (records of distribution of electric charges on a screen, or circulating sound pulses in mercury-filled tubes) provide a better model for memory, not because they are self-renewing—for mechanical devices are also theoretically capable of that—but because they do not tie down discrete bits of information to definite locations, just as discrete memories are not attached to definite neurones or groups of neurones. Some are inclined to favour the model of distributed electric charges; others point out that since the anatomy of the brain does reveal closed neural circuits, the model of circulating impulses is a better one; the latter idea, that 'reverberating chains' of neurones might be storing our memories has lately both gained publicity and popularity, and engendered strong opposition[11].

The new suggestions about the character of memory traces have not been put to a serious empirical test. Indeed it is difficult to see how to test them. Nevertheless a fresh line of research and study has been suggested. Whether it will ultimately prove to be a fruitful one, remains to be seen. The orthodox outlook and approach in the search of the engram has largely failed; the new analogies if not taken too literally are not so absurd as to be dismissed outright.

Related to the problem of the character of memory traces is a further puzzle: what precisely happens when a present stimulus impinges upon a memory trace, that is, when recognition occurs? Simple but rather inadequate models of recognition present no great difficulty.[3] A more sophisticated approach to the problem is that favoured by a group of workers in Chicago. Theoretical mechanisms capable of recognizing universals and *Gestalten* and, therefore, simulating an important aspect of human recognition have been described.[12] What is the outcome of this type of work? On the one hand, there is the intrinsically fascinating intellectual pursuit, which, like so many other theoretical studies in, for example, the field of mathematics may or may not turn out to be of practical value. On the other hand, there is the implication that mental functions are in principle capable of being explained in mechanistic terms; how one might regard such metaphysical implications of Cybernetics has been a subject of discussion elsewhere in this article.

At the purely psychological level, it might be asked what the model of memory trace must be like to satisfy the conditions according to which memories fade and undergo change. One model has been used which embodied a binary input and an electronic storage system. It was not suggested that either the structure or the function of these arrangements paralleled the nervous system. Nevertheless the behaviour of the 'set-up' as viewed by an outside observer would resemble in some ways that of human memory. It is likely that models using simpler components and other types of input could be constructed to behave in the required manner.

6. PERCEPTION

The scansion theory of the mechanism of perception, though put forward independently[13] has to a considerable extent been associated with cybernetics. It is suggested that incoming sensory signals are scanned by a steady oscillatory activity after the manner that a picture is scanned by the electron beam of a television camera. In this the carrier frequency is modulated by the incoming signal. It is further suggested that the alpha-rhythm as it is known in encephalography, constitutes the scanning mechanism of the brain. Its presence indicates readiness to perceive. It disappears when perception begins to take place. It is pointed out that Nature, like telecommunication engineering, would be expected to adopt this method of registering messages because it is an economical one in the number of communication channels used.

The present status of the scansion hypothesis is somewhat doubtful. But the explanation of some features of perception offered by this hypothesis is interesting and almost certainly amenable to further testing.

At any rate, this novel way of thinking about perception is

not so odd as might appear at first sight. It is in some measure in keeping with what we know to be the mode of functioning of other parts of the nervous system. Any movement made by the body is made against the background of muscular tonus. Any particular bodily movement leads to a re-patterning of the constantly present stream of impulses; it amounts, therefore to the modulation of the basic pattern of impulses. Alpha-rhythm has been called the cortical tonus. Like muscular tonus, it represents a normal unstrained condition of readiness or vigilance. It is demonstrably disturbed by perception just as the tonus condition is disturbed by bending, or sitting down, or walking. The importance of this borderline field between physiology and psychology cannot, of course at this stage be appraised. The use of the mechanical and electrical analogies in the study of perception, although it cannot be regarded with great enthusiasm, ought not to be rejected out of hand. Any ideas which seem plausible ought to be scrutinized with care in this difficult field.

7. SUMMARY

The influence of Cybernetics may be considered under three headings:

(1) It has been suggested that the positive contribution of Cybernetics is a hypothesis which neurophysiologists may test; namely, that negative feed-back mechanisms underlie the working of the central nervous system. If it is discovered that servo-mechanisms play an important part in the working of the brain, this fact will not be a revolutionary one. Neurophysiologists have been interested in discovering such subtle mechanisms for some time. What they want is to discover what mechanisms operate and with what results in specific cases where a particular group of nerves or cells are concerned. It may be the case that the construction of an electronic analogue might suggest how a particular part of our nervous system operates, but it seems doubtful if any model or set of models can adequately represent the system as it functions as a complex whole. The only test is the pragmatic test: try to use Cybernetics as an aid to Neurophysiology and see whether the results are useful.

(2) Some cybernetic models possibly suggest study projects within the psychology of learning and the psychology of thinking. Otherwise the main contributions to general psychology are analogies drawn from robot simulations of perception and memory

which do not at present appear capable of yielding testable hypotheses. The construction of cybernetic models may have some limited use in the field of experimental psychology.

(3) Many of the questions raised by Cybernetics are metaphysical. The present time is not exactly a confident period in the history of metaphysical thinking. We are not certain what results, if any, we can reasonably expect metaphysics to achieve. Some philosophers regard any form of metaphysical thinking as worthless: others assert that the proper aims of the traditional metaphysics have been transformed and may be successfully accomplished by the new techniques arising out of contemporary logic and semantics. Others attempt to carry on the old traditions in a modern idiom. What can one say about the metaphysical practices of cyberneticians? One interpretation of their activities as metaphysicians seems reasonable to the authors. Metaphysical systems, as constructed in traditional philosophy, often serve the function of expressing a moral outlook. Moral attitudes, sentiments, interests and needs are rationalized and related to other aspects of human thought. Thus Cybernetics develops a mechanistic-materialist philosophy of mind as a support to scientific materialism; man is regarded as something analogous to a highly complicated robot.

There are some people who are shocked by this attitude and insist that man is an immortal spirit loved by God and tempted away from God by the Devil; and there are yet others who do not believe in unseen spirits but who firmly reject the notion that living things are a special case of mechanical things. A study of anthropology and comparative religion gives us, as we all know, an interesting variety of such 'views' which are held, or have been held, by different groups of men. Some people become excited in the support of one or other of these attitudes; other people remain cold and apathetic to their attractions. Philosophers used to engage in the construction or alteration of such systems. Today they tend to be neither excited nor apathetic in their attitude towards them. Instead they attempt to show what people are implying when they discuss metaphysical problems and attempt to convince each other of the truth of one particular view. Thus, criticism of metaphysical thinking is largely a matter of logic and semantics, although, as Professor J. T. D. Wisdom has said, we sometimes need to call on the psychologist to reveal the hidden sources of a particular attitude and its expression in metaphysical utterance.

One function of Cybernetics, in its metaphysical aspect, seems to be that of bringing together ideas from different, though related, fields of study. This service as a 'melting-pot' provides new sources of mutual stimulation for philosophers, psychologists, neurophysiologists, psychiatrists and others concerned with the 'understanding of the mind', and counteracts the danger of narrow over-specialization. The entire field is essentially governed by trial-and-error methods of an analogical kind which belong to a pre-scientific stage in the investigation of phenomena.

The stage of enquiry is important and yet has received inadequate attention from logicians. It may be that the construction of analogies in Cybernetics will suggest what are the criteria governing reasonable conjecture.

REFERENCES

[1] Ashby, W.R. (1952). *Design for a Brain*. London, Chapman and Hall.

[2] Thomson, R. and Sluckin, W. (1953). *Br. J. Phil. Sci. 14*, 138.

[3] Tustin, A. (1953). 'Do Modern Mechanisms Help us to Understand the Mind?' *Br. J. Psychol. 24*, 24-37.

[4] Mace, C.A. (1948). 'Some Implications of Analytical Behaviourism'. *Proc. Aris. Soc., N.S. 49*, 1-16.

[5] Farrell, B.A. (1950).'Experience', *Mind 52*, 170-198.

[6] Mackay, D.M. (1951). 'In Search of Basic Symbols' in *Cybernetics* (ed. von Foerster), New York; Macy Foundation.

[7] Walter, W. Grey (1953) *The Living Brain*, London; Duckworth.

[8] Howard, I.P. (1953) 'A Note on the Design of an Electro-Mechanical Maze Runner', *Durham Univ. Res. Rev.* No. 4, 54-61.

[9] Shannon, C.E. (1951) in 'Cybernetics' (ed. von Foerster) New York; Macy Foundation.

[10] Lashley, K.S. (1950). 'In Search of the Engram'. *Symp. Soc. Expl. Biol.*, No. 4 (*Physiological Mechanisms in Animal Behaviour*). Cambridge.

[11] Jeffress, L.A. (ed.) (1951) *Cerebral Mechanisms in Behaviour* (The Hixon Symposium). New York; Wiley.

[12] Pitts, W. and McCulloch, W.S. (1947) 'How do we know Universals', *Bull. math. Biophys.*, 9.

[13] Craik, K.J.W. (1943). *The Nature of Explanation*. Cambridge Univ. Press. (see also Grey).

5

MODELS AND THE LOCALIZATION OF FUNCTION IN THE CENTRAL NERVOUS SYSTEM

Teleology is an old trap, not a particularly dangerous one these days, but one which caused much confusion in the nineteenth century when psychological problems were first put to systematic laboratory study. Nowadays we seldom, if ever, attempt to 'explain' natural or even mechanical phenomena by ascribing to them a special 'purpose'; in fact one current tendency is to look at known mechanical systems, and try to extrapolate their principles to the study of mysterious living systems. Often this can be tremendously useful, as Richard Gregory points out in this contribution, drawn from the famous National Physical Laboratory Symposium on 'The Mechanization of Thought Processes'. To find out that the heart is a pump is a great discovery, and one with immense practical value. But how do we distinguish between the functionally important and the accidental aspects of complex systems? For example, if a motor car engine is believed to be a hair drier, the exhaust has obvious functional value, but why the noise, the smell, and the rotating shaft? Once the true nature of the shaft is discovered however, the noise, smell, heat, etc. are realized as being 'accidental' appendages of what is now seen to be a remarkably efficient system. In this paper, Gregory, who is an experimental psychologist with an engineering bias, considers the important question; 'what sort of explanations are appropriate for biology?', and deals in particular with models of localization of function in the cerebral cortex.

Further Reading
Lashley, K. S. (1966). 'In Search of the Engram'. Reprinted in *Key Papers; Brain Physiology and Psychology*. Butterworths; London, and University of California Press; Los Angeles.

Scholl, D. A. (1957). *The Organization of the Cerebral Cortex* London; Methuen.

Reproduced from *N.P.L. Symposium No. 10* (1959) 671-681 Crown Copyright reserved, by courtesy of The Controller, H.S.M.O.

MODELS AND THE LOCALIZATION OF FUNCTION IN THE CENTRAL NERVOUS SYSTEM

R. L. Gregory

SUMMARY

The general question is raised: what sort of explanations are appropriate for biology? Teleology is rejected, while explanation in terms of conceptual models of the engineering kind is accepted. It is argued, however, that the engineering approach does not rid us of the necessity for making decisions on the purpose of observed structures or behaviour. Engineers generally know the design ends of their devices, but we have to guess the functional significance of biological systems. Certain difficulties are discussed. These considerations lead to the problem of distinguishing between 'accidental' and 'functionally important' features of biological systems. The criteria for distinguishing between these would seem to require conceptual models.

Localization of function is discussed in general terms. The use of the words 'localization' and 'function' is discussed, and a comparison is drawn between removing or stimulating parts of machines and brains. It is argued that the performance changes associated with removing parts of machines cannot be understood except in terms of functional models of the machine. The same would seem true for the central nervous system. In the case of independent parallel systems, such as telephone installations, much might be learned given only the crudest 'anatomical' model, and this is largely true of the peripheral nervous system. In the case of the cortex, this approach is likely to be misleading, especially if it is a tightly coupled system, for then removal of part of the system will either have little effect, except in certain conditions, or will introduce new functional features, for we now have a different system. These new features can only be understood, and only have significance, in terms of a functional model. It is concluded that ablation and stimulation of the brain may, and indeed does, bring out interesting facts, but that these must be interpreted in terms of models, for without a model we cannot say what is localized.

INTRODUCTION

When a biologist considers the fundamental question: 'What *is* an explanation of behaviour?' he may have a sense of conflict, even of paradox. Biology seems to be a science in its own right, or set of sciences having common aims, and so it should have its own language and explanatory concepts; yet when any specifically biological concept is suggested and used as an explanatory concept it seems to be unsatisfactory and even mystical. There are many biological concepts of this kind: Purpose, Drive, *elan vital*, Entelechy, Gestalten.* Physicists and engineers seem, on the other hand, to have clearly defined concepts having great power within biology. Why should this be? Is it that biology is not sufficiently advanced as an explanatory science to have developed its conceptual systems sufficiently far? Or is it that biologists should not look for *special* concepts, uniquely applicable to living systems? This latter view implies that biologists should think of living systems as being examples of physical or engineering systems in which case engineering language should be applicable for description and explanation in biology.

If an engineer is presented with a piece of equipment about which he knows very little, he will at least know that it was designed by men for some purpose he could comprehend. He is unlikely to find any quite new (or any highly inefficient) techniques involved in its design or construction. The 'black box' situation is artificial in engineering: the boxes an engineer sees are not as black as all that. In the case where the box *is* sealed (perhaps as a game, or as a test, or in a war), he still has a lot of information about it if he only knows that it was made for human beings by human beings. Biological systems are not designed by humans and the purpose of many of their characteristics may be highly obscure. Thus in important respects the engineer is not on his home ground when he advises the biologist.

*Some biological concepts have been given 'engineering' definitions; for example drive and purpose. Whether they still cover the cases required by biologists is an important question. Perhaps Survival of the Fittest is a specifically biological idea, and in a sense an explanatory concept, though it does not attempt to explain the functioning of an organism but only how it came to be 'designed'. Although the idea started as a biological theory, it may be extended to other cases, in economics for example, and it may be expressed in non-biological, largely statistical, language. Thus, it is not specifically biological.

Biologists are shy of the notion of purpose; in particular, they reject teleological explanations as 'unscientific', It is not clear however, that classification of biological structure and function in terms of purpose should, or can, be avoided. At first sight engineering analogies may seem to do this, but this is not so. To take an example of Dr. Uttley's, it is a discovery, and a most important one, that the heart is a pump. It would be impossible to understand it without recognizing this. Biological systems are adapted, through genetic experience, to serve ends which we might, or might not, discern. We have to discover what this end is before we can think up what sort of system it is, or assess its efficiency.

At this point we should try to clear up an old bogey. We can see what the purpose of a piston or a boiler is, once we understand steam engines, and we can see the functional purpose of the heart, or the eye. Why, then, do we not find purpose in inanimate nature? It is poetic, and most misleading, to say, for example, that the purpose of rain is to water flowers, and yet one might say that a purpose of rootlets is to suck up moisture. Is this because rootlets are alive, but rain is inanimate? Surely not. Rain does not fall especially on flowers, but flowers do take advantage of what rain there is, by specially adapted structures. Only living things and machines are adapted, in this sense, to take advantage of their environment, and only they display purpose. To state a purpose is, evidently, to specify what a thing is adapted or designed to do and we only find adaption, or design, in organisms and machines. If one supposed a car engine to be a hair drier the exhaust pipe would have an obvious use, but the rotating shaft would not. It would seem to be a very poorly designed hair drier; noisy, smelly, dirty and inefficient, with a most annoying rotating shaft. The noise, and smell and the heat will appear 'accidental' once the shaft output is recognized as such, and quite different estimates of its efficiency will be made.

It is often difficult to decide whether an observed feature is functionally important or merely accidental. Any physical system will have some colour, but generally this is unimportant to its function. It requires a knowledge of function to classify observed features into 'essential' and 'accidental' properties, yet without hypotheses of function this classification is impossible. The engineers' insights into the functional significance of observed features would thus seem of the greatest importance to the classification of biological observations and findings. Biologists have tended to make this distinction implicitly, as may be shown by looking at their choice of words. Let us consider an example of this. *'Fatigue'* indicates loss of efficiency after prolonged use; *'Adaptation'* suggests that a change which might be identical, and might occur under exactly the same conditions as fatigue, is useful. Thus, 'the retina has recovered from *'fatigue'* and, 'the retina has become *adapted'*, may refer to the same facts, and yet these sentences have different meanings for they imply different decisions on the significance of the facts. A useful feature in a machine might be regarded as having been specially designed to that end. Useful biological features tend to be regarded as the 'essential' features of the system given by specially adapted mechanisms. The decision between 'functionally important' and 'accidental' is often difficult to make in biology. Consider the continual small tremors of the eye ball. Are these useful, or do they represent a failure of precise control? Do they serve to aid vision, perhaps by preventing retinal fatigue at contours by producing re-stimulation of the retinal cells? If the tremor can be shown to be useful in this respect (and this is an empirical question open to experimental test), would it be supposed that there are special mechanisms evolved to produce the tremor, as there are mechanisms which have evolved to give accommodation changes, and colour vision? Would it be worth looking for such special mechanisms? If we found a bit of the brain which seemed to serve no function but to induce tremor in the eye, would we be more inclined to think that tremor has been specially developed? If we took this course, perhaps we should look at earlier eyes, for the tremor mechanism might *have* been useful though now it might not be. Suppose in fact the tremor *is* accidental, perhaps mere noise in the system; a biologist might still do well to study its effects on vision, but its status would affect his way of thinking about the visual system as a whole. It might affect the choice of explanatory model we would develop. The engineer may be able to help here by suggesting what special features of the phenomenon to look for to make the decision between 'essential' and 'accidental'. Thus hunting might be distinguished by statistical or other criteria, from noise. This sort of suggestion would seem to be of the greatest benefit to biologists. The engineer can help not only in suggesting models, but perhaps even more important, by suggesting what to look for to develop criteria for deciding between 'accidental' and specially adapted or 'essential' features.

Within the field of behaviour, probably the most important decision of this kind is whether the slowness of learning is specially adapted or is a limitation of the brain mechanism. One-trial learning is rather rare; why is this? If we knew what takes place in the brain during learning, then we could probably say whether or not this is a weakness of the design of the system, or whether it is specially built-in according to genetic experience. Without this knowledge of the mechanisms subserving learning we can only ask: would one-trial learning be, on the whole, an advantage? If we decide that it would be a good thing, then it will appear a weakness of 'design', or inadequacy of the materials of the brain, that it is so rare. This is a shaky argument, but it is frequently used in biology for lack of anything better. In fact, we can point to advantages in *not* having one-trial learning; for unreliable inductive generalizations are less likely to occur. It is notable that Pavlov required a silent laboratory, with nothing untoward occurring during his Conditioning experiments: distractions made the process longer.* We might suggest that slow learning is generally useful, we may then be tempted to think that the brain could perfectly well have provided fast learning but it would have been inappropriate. If so, certain types of engineering models may be ruled out. The biologist's assessment of the utility of the things he observed may affect the choice of 'engineering' model. The engineer might well question the biologist, and ask whether there are *any* cases where, for example, very fast learning does occur. If he can find such cases the limits of the system will have been extended. This might open up further possibilities, or rule out others. 'Imprinting' is interesting in this connection, for here very fast learning does occur, and it is plainly useful.

ON HOW AN ENGINEER MIGHT LOOK AT THE NOTION OF LOCALIZATION OF CORTICAL FUNCTION

1. *What Neurologists mean by 'Localization of Function'*

It is often suggested by neurologists that some function is localized in a more or less specific region of the central nervous system. Thus it may be suggested that speech, or some particular feature of speech behaviour, is localized in Broca's area, in the left frontal lobe. Or it may be said that the pre-frontal gyrus is a

---

*This is a signal/noise situation. The irrelevent events constitute noise which has to be ignored as far as possible if responses are to be appropriate. This is a very general problem for organisms making decisions on sensory or other data cf. Gregory[3,4], Barlow[1,2].

'motor area'. It is not always clear what neurologists mean by saying that a function is localized in a specified region of the brain. If this is not clear, any test or experiment involving behaviour is made difficult, and discussion is likely to be confused. I shall now try to examine the language used to describe localization of function. We may then see whether the techniques used by engineers for establishing and describing function might be helpful in relating behaviour with neural systems.

If we consider the phrase 'localization of (cortical) function', we can be puzzled by the use both of 'localization' and of 'function'. The neurologists generally means, evidently, that *a given region (of the brain) is especially associated with some particular type of behaviour*. This seems to be clear from the contexts in which the phrase is used—mainly ablation and stimulation experiments in which behaviour changes are correlated with brain damage or stimulation. This idea that a given area, or region, of the brain is particularly associated with certain behavioural functions goes back at least to the phrenologists. They spoke of a 'bump of intelligence' where, in some sense, intelligence was supposed to reside. The bigger the 'bump' the greater the intelligence.

2. *'Localization'*

At first sight, at least, it may seem that the neurologist speaks of regions of the brain in the same way that the Geographer speaks of regions of the Earth. Thus: 'The Sahara is a region in Africa' and, 'the striate area is a region of the occipital lobes', are similar sorts of statement. A desert ends when the sand becomes soil: the striate area ends when the cell structure changes in a recognizable way. Suppose now that there is no visible dividing line to delimit 'functional' regions of the brain: How could we specify such regions? The same thing happens in geography, and the analogy may prove helpful. In *physical* geography natural features, such as seas or continents, are bounded by shores; deserts end where the climate and soil change. In *economic* geography countries, or trading areas, they may not be so bounded: they are essentially *functional* areas. These are functional boundaries, and make sense only in terms of the life of the people in each country—their customs, economics and government. These things cannot be observed like shore lines, by simply looking. A further point—a country is usually fairly compact, for reasons of transport and communication, but this may not be so. Thus in a sense the British Empire might be said to be, or to have been, one country though scattered across the world. The bits were linked by connections which could be described as 'functional' but they could not be observed as, at any rate static, 'structure'.

The distinction made between physical and economic geography holds for the neurology of the brain. A region may be isolated and named either (a) by virtue of its special *structure*, or (b) by virtue of its special *function*.

3. 'Function'

In normal speech, to specify an object's function is to say what causal part it plays in the attainment of some end. It would seem that an object's function may be specified in two quite different ways, and the difference between these is important. Consider a rheostat used as a dimmer to control the brightness of a lamp. The resistor may be referred to in terms of its property changing the current in the circuit, or simply as a dimmer. The word can only be used in the first way within the context of electrical theory; it is essentially a *technical sense* of the word. The same component may also be referred to quite non-technically, as; 'the thing which dims the lamp'. Here no special knowledge is required, beyond the generalization that when the knob is turned the lamp dims. The grounds for using the word 'function' are quite different in the two cases, depending upon whether it is used to designate an essentially *inductively* derived generalization that *A* produces *B* (turning the knob has always dimmed the lamp, and so it is a dimmer) or whether it denotes a *deductive* inference within a theory: (added resistance reduces the current flowing in a circuit; the reduced current gives less power.... therefore the lamp must dim). If the current was not reduced, it could not *possibly be* a resistance. It should be noted that we may be mistaken in *calling* a given object a resistance, but if it *is* a resistance then it *must* have certain functional properties in a given system. Now it is clear that this second, deductive, use of the word can only be used within a deductive system, or model. The burning question is: has the neurologist got such systems or models to enable him to use 'functional' in this technical way?

It seems clear that the word 'function' is generally used in neurology in the non-theory laden, inductive, sense. If it is said: 'speech is localized in Broca's area', or 'the function of Broca's area is to (with the vocal chords and many other things) produce

97

speech', then we are using an inductive argument. The evidence for the induction is based largely on speech defects which are observed with lesions in this region of the brain. Broca's area is said to be a speech centre, without a notion of just *what* it does or *how* it does it. Again, it may be said that the striate area is used for vision, but we have no idea what processes are going on in this region during visual perception.*

It often happens that with man-made machines we can recognize the various components for what they are, or at least some of them. We may see that there are, for example, motors, resistances, levers or bearings in the machine. Once the components are identified as functional units, then deductive inference is possible. Of course we may always be wrong at the identification stage of the process: a resistance may be mistaken for a condenser, and then the premiss for the argument will be wrong. But the argument is still truly deductive, though the conclusion may then be wrong. Now if nerve cells varied in form in more different ways, each type having corresponding functional properties, then the neurologist would be in the position of the engineer in having readily identifiable components from which he might infer the function of the circuit. Unfortunately, the neurologist is perhaps never in this position, for the brain is like porridge. This means that this way is generally not open, for we cannot, at least at present, identify and recognize function from appearance by looking at cells and their connections.

If any change takes place upon the removal of part of the brain, the changes are either (*a*) *loss* of some feature of behaviour, or diminution or worsening of some skills, or (*b*) introduction of some *new* behavioural features. Now it is often argued that if some part of behaviour is lost, or diminished in efficiency, then in some sense this behaviour (or rather the

---

*The neurologist's concern with cortical function is the extreme case: it would be interesting to consider questions of localization of function in simple cases. There are, however, comparatively simple examples to be found in the brain: thus certain regions can be described as 'chemo-receptors', serving to monitor the blood concentration of CO_2. The signals from these regions have a direct action on respiration. I do not question that some areas have specific functions, but to state what their function is we must have a model of the system within which they play their part. To name an area as a 'chemo-receptor' is to give it a (sensory) function.

It might be recognized as such either (*a*) by the recognition of distinctive properties (like recognizing a rose, or a friend) or (*b*) by recognizing its place in a system. (*a*) Requires prior knowledge of such components or parts, (*b*) requires a knowledge of the rest of the system in functional terms. In either case, to specify the function of part of a system we must know something of *what* it does and *how* it does it. This is true of the simplest, as well as the most complex, types of system.

causal mechanisms necessary for this behaviour), are localized in the affected region. But does this follow? We may assume that the association is no chance association but a causal one, and still seriously doubt whether before the advent of the lesion the region in question contained causally necessary mechanisms for the affected behaviour. To illustrate this, we can take an example from radio engineering. If a main smoothing condenser breaks down, shorting the H.T. to earth through a low resistance, the set may stop working or work in a peculiar manner. The oscillator may stop although the rest of the system may continue working normally. Would we then say that the condenser was functionally important for the radio's oscillator stage, but not for the rest? If so, we would be wrong. Its purpose in the system is to smooth the ripple for the whole system, but it happens that a part of the system is more sensitive to reduction in a supply voltage than the rest. Suppose that when the condenser breaks down, the set emits piercing howls. Do we argue that the normal function of the condenser is to inhibit howling? Surely not. The condenser's abnormally low resistance has changed the system as a whole, and the new system may exhibit new properties, in this case howling.

There are physical systems where removal of a part removes a specific feature of the output. Thus consider a piano: if a string, a hammer, or a key is removed one note is lost, and the rest may play as before. A piano is largely an arrangement of independent parallel systems, each with its own input (a key) and output (a string). Here functional localization is a simple matter—a piano tuner soon knows where to look for any trouble. This is not in general true of machines, where loss of a part may produce the most bizarre symptoms. Only by understanding the principles involved, and the causal functions of the parts, can the trouble be explained. Further, even where there are semi-independent systems, a fault in a part serving the various parallel systems may have a selective effect on them, and this can be confusing. Thus, reduced air pressure on an organ might affect some pipes a great deal and others not at all.

Mr. P. E. K. Donaldson has suggested to me that this point applies particularly to tightly coupled systems, and the brain would seem to be such a system.

4. Arguments for Stimulation

If a part is stimulated and something happens, such as a

muscle twitching, or the patient reporting a cloud of balloons floating over his head, what can be inferred? The considerations here are similar to the above. We now have a different system which might have quite new properties.

The functional 'centres' of Hess, and the ethologists are interesting. If stimulation of the given region produces a sequence of movements similar to, or identical with, an observed innate behaviour pattern then, it is suggested, this region is the locus or 'centre', of this behaviour pattern. But there are difficulties in this idea of a localized functional centre. The word 'centre' here suggests that the causal neural processes leading to the innate behaviour pattern are located in space, closely packed, and even that there are not other causal mechanisms in the same region. But, just as many races may live in the same country, perhaps talking the same language and perhaps not, so causal mechanisms subserving different end results could well reside entangled. They may not be grouped neatly but might perfectly well be strung about in tenuous filaments, very difficult to find, identify, or stimulate in a controlled manner. Perhaps most important, why should the stimulation by some arbitrary 'signal' leading to familiar behaviour patterns be regarded as more interesting than some bizarre behaviour? Surely all it could mean is that some organized set of causal sequences has been set off, leading to familiar behaviour; but this happens with lights shone in the eye, or sounds applied to the ear, or any other stimulus, yet no one wants to say that the causal mechanism, the 'centres' are in the eye or the ear, except in very special cases. Indeed, it would be surprising if stimulation of the cortex did *not* sometimes produce behaviour sequences, but it would not follow that the region stimulated contained the causal mechanisms responsible for the activity. Certainly it must under these conditions have something to do with the activity, but it may be far removed, both spatially and causally. One might suspect that if a complete normal behaviour pattern is elicited with an artificial stimulus, then the stimulated area is rather less likely to be directly responsible than if the behaviour patterns are bizarre, for the stimulus might be expected to produce disruption of normal function if introduced in the middle of a causal mechanism, even if the stimulus is the correct trigger for the mechanism.

'Localization of function' takes on further difficulties when we go from innate to learned mechanisms, for here each individual will differ according to what is 'stored' and also, perhaps, according to different developed 'strategies' adopted for thinking.

100

If the arguments (or at any rate the conclusions) suggested here are substantially correct, the changes produced by ablation or stimulation should be interpreted in terms of the changes to be expected in systems of various kinds. In a tightly coupled system it is impossible to specify a flow of information, or to say where any particular function is localized, for there is interaction throughout the system. Its performance may change when bits are removed, or stimulated, but the changes can only be understood in terms of the functional organization of the whole system. In short, we need a model to interpret the changes, and this is where the engineer should be able to help. The changes in behaviour associated with lesions might be important evidence for suggesting and testing models, and the models should serve to explain the effects.

Modern neurology started when large telephone exchanges were first being built. The 'telephone exchange analogy' has no doubt been useful in the study of the peripheral nervous system. but it is extremely misleading for tightly coupled systems, as the brain would seem to be. The input and output regions of the brain—the 'projection areas'—would be expected to be revealed fairly simply by suitable lesions or direct stimulation, for no doubt they lie in telephone-like pathways, as for the peripheral nerves, though even here there tend to be some cross connections which complicate the picture. The problem becomes acute where the nervous system is analysing or computing for here the system cannot be like a telephone exchange. Histological examination suggests more or less random networks in large parts of the cortex (Sholl[7]). Such systems are only beginning to be studied. Lashley's 'Equipotentiality' (that any part of the cortex is equivalent functionally to any other part, certain areas excepted) which he established for the rat brain (Lashley[5,6]) would be expected for such a system. If part of a tightly coupled system is destroyed there is generally very little effect on its performance, except when it is pushed to its limits, for it is now a smaller system. This is just what Lashley found. It is interesting that Uttley's conditional probability computer consists in part of random circuits; some destruction here would not have any devastating effect on the function of this machine. (Uttley[8]).

The behavioural effects of brain damage are of vital importance to the brain surgeon quite apart from their interpretations. They also provide fascinating data for insights into the function of the normal brain but only, I submit, if we look at the changes in the light of conceptual models of brain function. We only come to understand fully how car engines and radio sets work, even after reading about them, by noting their eccentricities and their ailments. We see why a certain 'symptom' is produced when we know how the normal system works, and we come to understand more fully how it works when we see and think about the symptoms of overheating, weak mixture, or whatever it may be. We could not possibly say what a plug in a car engine does simply by noting what happens when we remove it, unless we understand the general principles of internal combustion engines —we must have a model. The biologist has no 'Maker's Manuals', or any clear idea of what the purpose of many of the 'devices' he studies may be. He must guess the purpose, and put up for testing likely looking hypotheses of how it may function. In both these tasks, he should receive valuable help from the sympathetic engineer.

I wish to thank Mr. A. J. Watson, Dr. L. Weiskrantz and Professor O. L. Zangwill for valuable discussions, though I should not wish to commit them to the views expressed. I am indebted to Professor Zangwill for any knowledge I possess on these problems.

REFERENCES

[1] Barlow, H. B. (1956). 'Retinal Noise and Absolute Threshold'. *J. opt. Soc. Am.*, *46*, 634.

[2] Barlow, H. B. (1956). 'Incremental Thresholds at Low Intensities as Signal Noise Discriminations'. *J. Physiol., Lond*, *136*, 469.

[3] Gregory, R. L. (1955). 'A Note on Summation Time of the Eye indictated by Signal Noise Discrimination'. *Q. Jl. exp. Psychol.*, 7, 147.

[4] Gregory, R. L. (1955). 'An experimental treatment of vision as an information source and noisy channel. In *Information Theory:* Third London Symposium. Ed. by Colin Cherry. London: Methuen.

[5] Lashley, K. S. (1929). 'Brain Mechanisms and Intelligence', Chicago: University Press.

[6] Lashley, K. S. (1950). 'In Search of the Engram, Symposia of the Society for Experimental Biology IV. England; Cambridge University Press.

[7] Sholl, D. A. (1957). 'The Organisation of the Cerebral Cortex'. London; Methuen.

[8] Uttley, A. M. (1955). 'The Conditional Probability of Signals in the Nervous System' *R.R.E. Memorandum No. 1100*.

6

MACHINES THAT THINK AND WANT

The name of the author of this paper is one which the student of cybernetics will keep encountering, and he will sooner or later need to read several and not just one of his papers. Professor Warren McCulloch (as he states at the beginning of his article) started off as a philosopher, and with considerable optimism looked to psychology for answers to certain critical questions. Understandably disappointed with the lack of progress in psychology at the time, he then moved hopefully into physiology. The hybrid scientific approach which resulted put McCulloch in the forefront of the pioneers of Cybernetics and led to some remarkable publications and, in particular, some outstanding collaborative work. 'Machines that think and want' is a provocative title for what is in fact a concise and fundamental essay—though one which is peppered with exciting ideas. McCulloch's papers are persistently optimistic in the belief that our brains and ultimately our behaviour are discoverable and comprehensible. At the same time there is the constant gentle warning that as scientists we should never take ourselves too seriously.

The references below are designed for those who feel they would like to extend their knowledge of McCulloch, rather than of brains in general. The first is an outstanding paper of its period; the second, a book of his selected works, including some of his slightly puzzling poetry.

Further Reading

McCulloch, W.S. and Pitts, W. (1947). 'How we know universals'.
  Bull. math. Biophys. 9, 127-147
McCulloch, W. S. (1965). *Embodiments of Mind.* MIT Press,
  Cambridge, Mass.

6

MACHINES THAT THINK AND WANT

Warren S. McCulloch

Some twenty-five years ago I left philosophy and turned toward psychology hoping to answer but two questions: First, how do we know anything about the world—either its particulars as we apprehend events or its universals as we know ideas? Second, how do we desire anything—either physically, as we want food and drink or a woman and a bed, or mentally, as we seek in music the resolution of a discord or, in mathematics the proof of a theorem? From psychology I turned to physiology for the go of the brain that does these things. Twenty ideas that came but one a year can be said in 20 min because they spring from traits of neurons in simple circuits.

Think of a neuron as a telegraphic relay which, tripped by a signal, emits another signal. To trip and to reset takes, say, a millisecond[25]. Its signal is a briefer electrical impulse whose effect depends only on conditions where it ends, not where it begins. One signal or several at once may trip a relay, and one may prevent another from so doing[26,27]. Of the molecular events of brains these signals are the atoms. Each goes or does not go. All any neuron can signal to the next is that it was tripped, but, because it signals only if tripped, its signal implies[32] that it was tripped. Thus the signal received is an atomic proposition. It is the least event that can be true or false[33]. If it is unnaturally evoked it is false, like the light you see when you press on your eyeball. How you find it false is another question[32].

Of two possible atomic events there are four cases, each shown by putting a dot into one of 4 places in an X—A happens, place dot on left; B, right; both, above; neither, below. Every next atomic event, depending on two present atomic events, is shown as a dotted X of which there are 16, being all the functions of the calculus of propositions. The first, with no dots, shows the signal of a neuron never tripped—call it contradiction; and

105

the last, with 4, the signal of one tripping itself every milli-second—call it tautology. Only the upper row of functions go in nets without tautology, for a dot below the X, showing the signal of a neuron neither of whose atomic propositions had come, is the signal of a neuron untripped. Give signals only two actions—one tripping, one preventing—and still nets of few neurons channel all these functions[33]. Moreover, we could replace nets of neurons modified by use by changeless nets of changeless neurons[33], and these functions do not depend on modes of tripping or preventing atomic propositions[8]—hence the purview of this calculus.

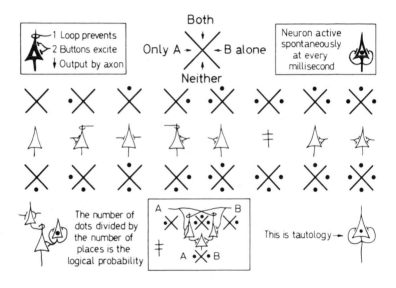

Figure 1.

To the psychologist this is most important, that such nets of neurons can compute any computable number[33].

The logical probability[51] of each of the 16 functions is the number of dots for separate signals divided by the number of possible places, which is 4. For the *a priori* probability that a neuron is in a particular state is $\frac{1}{2}$, i.e. 2^{-1}, and that n inde-pendent neurons are in a particular state is $\frac{1}{2}^{-n}$ i.e. 2^{-n}. One unit of information defines the state of one neuron, so n units, the state of n neurons—hence the amount of information is the logarithm of the reciprocal of the probability of the state. If in a particular millisecond we are given an X, since its probability

106

is ¼, its amount of information is 2 units. But the frequency of realizing X from independent signals is the product of their separate frequencies. Hence, averaged through time, X will transmit no more information than its afferent neurons singly. Similar considerations apply to all functions. So n independent neurons, regardless of their functions, transmit on the average n units of information per millisecond. Moreover, n neurons can be in any one of 2^n states in one millisecond and in 2^{nt}, in t milliseconds. So, in p milliseconds one neuron can transmit as much information as p neurons in one millisecond[33]. Information measures the order of an ensemble in the same sense that entropy measures its chaos[49]. In Wiener's phrase, information is negative entropy; and in our X's the more the dots the greater the entropy.

Neurons can be exchanged for milliseconds, and functions of greater for those of less logical probability, but we cannot get out of the brain more information than went into it. We express much less, because a nerve of a thousand axons has $2^{1,000}$ possible states—which is more than there are particles in the universe,—whereas the muscle it drives has a paltry thousand possible tensions.

On the receiving end we reject far more information. Consider the eye. Each of its hundred million rods and cones may at one moment fail to fire with much light, and at the next, fire with almost none[18]. If this be by chance it may be ignored in large groups of contemporaneous impulses. We detect the agreement of these signals by their coincidence on the ganglion cells and relay only that information to the brain. Thus we pay for certainty by foregoing information that fails to agree with other information. No machine man ever made uses so many parallel channels or demands so much coincidence as his own brain, and none is so likely to go right. Similarly, our hypotheses should be so improbable logically that, if they are instanced, they are probable empirically.

Because light falling on a rod may or may not start a signal, that signal implies—but only implies[48,51]—the light. If a bipolar cell is tripped by the rod's signal, then the bipolar's signal implies the rod's signal, and it, in turn, the light. Similarly a signal by a ganglion cell implies the bipolars, the rods, and so the light. Thus what goes on in our brains implies—but only implies—the world given in sensation. The domain of these implications extends only backward in time. Wherefore we know only the past. These bounds are proper to cognition. If each neuron were to signal when and only when it had received one of several signals, we could proceed in the opposite direction from brain to muscle, with every step an implication. The present would imply fate as well as fact. We could then deduce our deeds from our thoughts and our thoughts from our sensations. Rather this is the nature of intention; that it falls short of implication to the extent that aught intervening thwarts us. Perforce we discern that we will from that we shall do. Such bounds are proper to conation.

Now a signal anywhere implies an event at just one past moment unless there is a loop it may circuit[32]. Once started in a circle it implies an event at any past time which is a multiple of the period of the circuit. A set of such signals, patterned after some fact, as long as its figure endures, implies a fact of that form. This form-out-of-time is an eternal idea in that temporary memory which generates objects new in our world; for it recognizes them on first acquaintance[47]. But these fleeting figures people only the specious present, and when the brain rests[6] they are no more anywhere. The aged often have no other memory[23]; but in young brains, use leaves a trace whereby the ways that led us to our ends become ingrained. We must make and read our record in the world. This extends the circuits through our heads beyond our bodies, perhaps through other men. Records are not signals, but may start them again in the old figure. To them as well our calculus applies, enriched by ideas, or timeless objects, attributed to some or all events[33]; for all forms of memory other than reverberations are but their surrogates[32]. Man-made computers store information as arrays of jots and titles only to regenerate signals in artificial nets of relays and so recall the past. For this the right array must first be found, then recognized; and, when the information is in divers arrays, it must be reassembled, or, as we say, remembered. Thus, over all, the procedure remains regenerative—a positive feed-back.

Circuits with negative feed-back return a system toward an established state which is the end in and of the operation—its goal[5]. Reflexes left to their own devices are thus homeostatic[9]. Each, by its proper receptors, measures one parameter of the body—a temperature, pressure, position, or acceleration—and returns that parameter to its established value. Similarly, within the brain, impulses through the thalamus to the cortex, descending from its suppressor areas indirectly to the thalamus, inhibit it, and so govern the sensory input as to extract its figure from its intensity[14,17,2]. If such circuits are sensitive enough they oscillate at their natural frequency. Their response depends on the time the returning meet the incoming signals, and at some frequency of excitation they will resonate; but another circuit, responding to the derivative of excitation, will check that resonance.

Clerk Maxwell first computed the go of the governor (*Figure 2*) of a steam engine[29]. It had two balls hung from the top of a vertical shaft turned by the engine. When speed increased they flew out, pulling up one end of a lever on a movable pivot, and

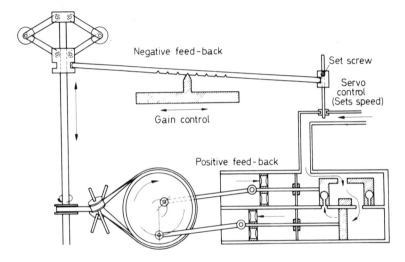

Figure 2.

so, by an adjustable link, throttled the engine. Moving the pivot changed the sensitivity, which we call the gain of the circuit; but changing the length of the link set the speed at which it was to run, regardless of load or head. Compared to the change in work the engine makes, that to set the link is negligible but its information all important, and we call this a servo-mechanism[30]. The stretch reflex is one in which the length the muscle will attain is selected by the brain for its own ends. Such government by inculcation of well-chosen values of ends sought by the governed works wonderfully well.

Purposive acts cease when they reach their ends. Only negative feed-back so behaves and only it can set the link of a governor to any purpose[40]. By it, we enjoy appetites, which, like records that extend memories, pass out of the body through the world and returning stop the internal eddies that sent them forth. As in reflexes, the goals of appetition are disparate and consequently incommensurable. They may be as incompatible as swallowing and breathing; and we are born with inhibitory links between the arcs of such reflexes. But of appetitions the dominance is rarely innate or complete, and we note the conflict whose outcome we call choice. When two physically or psychologically necessary acts are incompatible, 'God' cannot forgive us for not doing the one because we must do the other. The machine inevitably goes to hell.

Of any three circuits subserving appetition the dominance may be circular[31]. I have myself encountered this in aesthetics; for, of three very similar rectangles shown in paired comparison the first was consistently preferred to the second, the second to the third, and the third to the first, on the average and by the single subject. I discarded the data as inconsistent, whereas it bespoke consistency of a kind I had not dreamed of .We inherit from Plato a vain superstition called 'the common measure of all values'. These are not magnitudes of any single kind, but divers ends of divers circuits so interconnected as to secure dominance which, like as not, is circular. Economic arguments from curves of indifference and attempts to set up one scale for the strength of drives for food, water, sex, and whatnot are fantastic ways to beget a gratuitous headache.

Negative feed-backs run through a series of affairs until they come to their ends. We have mentioned the automatic volume-control of somaesthesia[2,14,17]. The pupillary reflex does the same for vision. An appetitive circuit can reduce whatever confronts it through a series of transformations to one so geared to our output that we cry 'Aha!' We transform the Pythagorean Theorem, by legitimate steps, to the axioms and postulates which are the canonical form of such theorems. Similarly, we translate an apparition anywhere in the visual field to the centre by turning our eyes in its direction. This reflex goes from the eye to the superior colliculus which, by double integration, determines the vector for the motor nuclei so as to turn the eye toward the centre of the apparition. The integral decreases to zero as the eyes come to rest[33]. The apparition is then in the canonical position whence to abstract its shape

No reflex can translate a chord to a canonical pitch; but pitches so project to Heschl's gyrus that octaves span equal lengths, and the information slants upward through scores of relays in vertical columns[1]. Let these be tripped only by coincidence of information with a second source of excitation—say the so-called alpha activity of cortex[3], now thought to arise from thalamic circuits[4,12,24,27,35,36] and, as this ten-per-second wave steps up and down through the cortex, its intersection with a given stream of information will move one way and the other along the axis of pitch. The output, descending vertically, then comes down in all points corresponding to these pitches in the output. When the information contains a chord of one pitch, the circuit has performed the nervous equivalent of a series of translations in pitch-conserving chord. Had we a negative feed-back to stop the alpha wave when its scansion had brought the chord

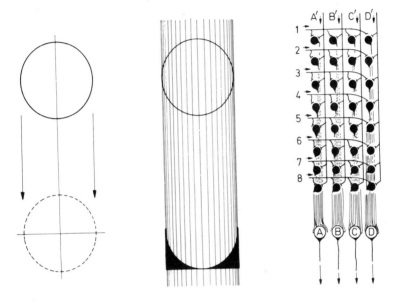

Figure 3.

to a canonical pitch, the process might end here, but there is no proof that it does.

A mathematically simpler way to secure an invariant is to perform successively on any apparition the whole group of necessary transformations, sum them termwise, and divide the sums by the number of transformations[33]. We need no division if the number of transformations is finite, as in brains. In *Figure 3*, the circuit at the right will translate a circle to the canonical position, as at the left, and sum over the translations, as in the centre, where the concave block indicates the number of signals to neurons situated at each point and transmissible over a single neuron if there is a circuit to scan the sum, appearing then in time as the variable frequency of the short lines beneath. Scanning exchanges one dimension in space for time[33], which makes it possible to house the required number of neurons in our heads.

111

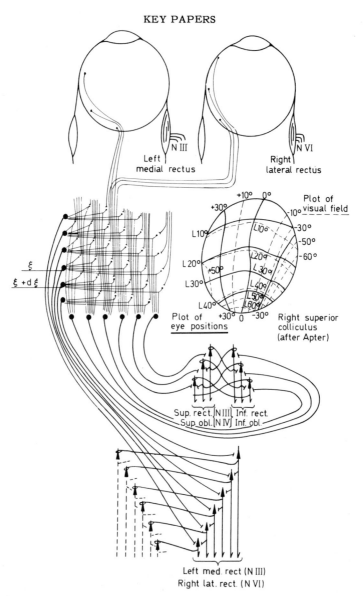

Figure 4. A simplified diagram showing occular afferents to left superior colliculus, where they are integrated anteroposteriorly and laterally and relayed to the motor nuclei of the eyes. A figure of the right superior colliculus mapped for visual and motor response by Apter is inserted. An inhibiting synapse is indicated as a loop about the apical dendrite. The threshold of all cells is taken to be one.

This diagram is used through the courtesy of the Bulletin of Mathematical Biophysics.

112

As a man's name written or spoken neither looks nor sounds like the man, so these invariants need bear no resemblance to the apparition save a 1-1 relation according to some projective law of denotation. Hence before we look we know not what to seek in the brain.

If we did not stop the sweep but summed the vertical columns, the primary auditory cortex would relay to the secondary auditory cortex a figure corresponding to the chord and not the pitch. Now local electric stimulation in Heschl's gyrus is reported as a tone;

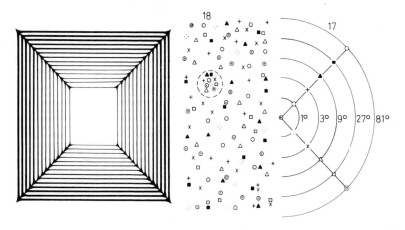

Figure 5.

whereas in the second auditory cortex, dreamlike experiences are induced[32]. All primary receptive cortices are columnar[45], which fits this theory of ideas. So does the sweep that disappears in sleep or anaesthesia, for its frequency is the greatest number of distinct perceptions per second occurring in any modality. Since a reflex (schematized in *Figure 4*) translates visual apparitions, the primary visual cortex needs only short columns. They are imbedded in branches of its incoming fibres which then ramify widely in the horizontal plane, turning toward the surface to end on cells increasing in size and decreasing in number from below upward[39]. Purely statistical considerations compel us to believe that large cells require summation from many endings to trip them. We would, therefore, expect the higher cells to respond to excitation over large areas, but not over small ones. A plane of excitation stepping up and down in such a cortex determines an output that runs through a group of contractions and dilatations of a figure in its input[33].

113

Now, the corners of an observed square tend to fill in the angles, producing an excitation like the figure on the left in *Figure 5*. These transforms are dominated by the diagonals. A pinhead spot of strychnine anywhere in the primary visual cortex, area 17, fires cells there in unison[13]. Their axons project at random to points in the secondary visual cortex[46], area 18, as I have indicated by scattering 18 replicas of each of the 11 signs placed along the diagonal lines in 17. Sheerly by chance there will arise in 18 some one small precinct, say that enclosed in the circle, to which the majority of these points project, and excitation will be maximal there. As the brain, going on coincidence, ignores or converts to inhibition[8] what falls short of summation, activity of the circled area in 18 implies the diagonals in the output of 17 and so a square regardless of size. Stimulation in area 17 yields but a vague ball of light fixed in the visual field[38]; whereas, in 18, such stimulation yields a well-defined form[38] of no particular size but fixed in space, for the eyes turn thither.

The action of circuits that average over groups is little affected by perturbation of input, of threshold, or even of synapsis if connections are only to cells in proper neighbourhoods[33] —a statistical order we may hope from our genes. Except possibly in the visual cortex, the horizontal connections in the grey matter need only be undisturbed for a few millimetres, for the bulk of synapsis is vertical[13,28], and recent notable experiments on the parietal cortex showed that a 2 mm grid of incisions through its whole thickness produced no loss of somaesthesia[43]. The visual cortex may have a radial structure to be injured by such gridding; but even large holes poked through it give little or no change in the place of maximal excitation in area 18, and although scotomata can be mapped in the visual field of a man so injured, he still sees forms which for him cross the scotomata, even as they do the blind spot of a normal eye. This continuity of perceptions is not given by physical or chemical fields[22], which can only corrupt information in an organ which is a mosaic of discrete units each capable of two discrete states. Per contra, it is a consequence of the statistical nature of the process of averaging over a group of transformations which could be equally well achieved were the particular relays physically far asunder. In vision, opposite sides of the vertical line through the fovea project each to the other hemisphere with no corresponding discontinuity of perceptual properties[19].

Closed paths probably mediate our predictions. Man-made devices start a shell toward that place in the air where the plane will be when the shell arrives and explodes. In these devices, series of events in time are autocorrelated to produce invariants, like kinematic routes, most likely to be run by things that have run some part of them. Such circuits tend to reverberate in us. Optokinetic nystagmus[7], which attempts to fix gaze on things in their passage in one direction, persists when they cease to move and it attributes to them a reverse motion without change of position. This is the only known predictive circuit we may assign to a particular part of the brain and even for this there are too many equally good hypotheses. Every scientific hypothesis is a prediction. We know only the past and little of that. Even if we are right in our superstition[21] that the previous state of the world does determine its subsequent state, any forecast we make may still go wrong—and a hypothesis forecasts the outcome of innumerable experiments. None can be proved true, any can be proved false. We must predict in order to live in a world where a false step means death. Here are errors at once inevitable and inexcusable, and the faith of any man who takes a single step is but an ignorance of ignorance concealed by universals extrapolated beyond fact.

Had we run down the fortune-telling circuits there would still be a gap in physiology which elsewhere dovetails neatly into psychology. We are ignorant of those alterations in our nets which underlie learning. The least hypothesis is that if impulses of one neuron C, just insufficient to trip a second, R, come concurrently with those from another, U, which can trip R, then, thereafter, C alone shall be able to trip R[41,42]. This gain in grip of C on R exemplifies the law of growth with use we now apply to other cells. Surely neurons are not immutable mummies under our microscope but are living beings, competing one with another for foot space on succeeding cells. The two axonal terminations that have been studied change with use. Myoneuronal junctions[10] and vagal endings in the ganglia in the frog's ventricle[15], on excess of stimulation, are wrecked but recover promptly and after more profound stimulation may reconstitute themselves among the useless fragments of their former state.

All protoplasm shows hysteresis[16], but that whereby anything becomes the symbol for something else—the conditional stimulus —is first found in the flatworm[34], the lowliest beast with a truly synaptic net[34]. It therefore behooves us to look to the synapse for the crucial change[10,15]. Mere change with use is not enough to substantiate the law of effect[44], which must depend on inverse feed-back. Therefore, our second hypothesis is that those combinations of cells whose activity was concurrent and which were last used when we came to our goals and the activity to its end become associated. The process of learning so conceived parallels item for item the magnetizing of a bar of iron. Had we the mathematics for either, it might do for both; for both are stochastic processes in which from chaos order evolves, each step fixing to some extent what was heretofore left to chance. The theoretical difficulty is that into the relations of pairs, there enter the relations of trios and into those of trios, those of foursomes and so on to infinity[50]. Such problems have not in the past proved insoluble. While I am waiting for the answer, I intend to seek the most transparent flatworms with the fewest neurons, teach them[44] and attempt to discern in the synaptic nets[34] under the microscope the difference between the scholar and the ignoramus.

REFERENCES

[1] Bailey, P., Bonin, G. Von, Garel, H. W., and McCulloch, W. S. (1963). 'Functional organization of temporal lobe of monkey (Macaca mulatta) and chimpanzee (Pan satyrus)' *J. Neurophysiol.*, 6, 121-128.

[2] Barker, S. H., and Gellhorn, E. (1947). 'Influence of suppressor areas on afferent impulses. *J. Neurophysiol.*, 10, 133-138.

[3] Berger, H. (1929) 'Ueber das Elektroenkephalogramm des Menschen'. *Arch. Psychiat. Nervkrankh.* 87, 527-570.

[4] Bishop, G. (1946) 'The interpretation of cortical potentials'. Cold Spring Harbor *Symp. quant. Biol.*, Cold Spring Harbor, 4, 305-319.

[5] Black, H. S. (1934) 'Stabilized feed-back amplifier'. *Electl. Engng.*, N.Y, 53, 114-120.

[6] Blake, H., Gerard, W. R., and Kleitman, N. (1939) 'Factors influencing brain potentials during sleep' *J. Neurophysiol.*, 2, 48-60.

[7] Borries, C. (1923) 'Reflektorischer Nystagmus'. *Msch Ohrenheilk. Lar-Rhino.* LVII: 547-570.

[8] Brooks, C. McC., and Eccles, J. C. (1947) 'An electrical hypothesis of central inhibition'. *Nature, Lond.* 159, 760-764.

[9] Cannon, W. B. (1929) 'Organization for physiological homeostasis'. *Physiol. Rev.* 9, 399-431.

[10] Carey, E. J. (1941) 'Experimental pleomorphism of motor nerve states as mode of functional protoplasmic movement', *Anat. Rec.*, 81, 393-413.

[11] Dempsey, E. W., and Morison, R. S. (1942) 'Production of rhythmically recurrent cortical potentials after localized thalamic stimulation'. *Am. J. Physiol.*, 135, 293-300.

[12] Dempsey, W., and Morison, R. S. (1943) 'Electrical Activity of Thalamocortical Relay System'. *Am. J. physiol.*, 138, 283-296.

[13] Dusser de Barenne, J. G., and McCulloch, W. S. (1936). 'Functional boundaries in the sensorimotor cortex of the monkey'. *Proc. Soc. exp. Biol. Med., 35,* 329-331.

[14] Dusser de Barenne, J. G., and McCulloch, W. S. (1938) 'Sensorimotor cortex, nucleus caudatus and thalamus opticus'. *J. Neurophysiol.,* 1, 364-377.

[15] Fedorow, B. G. (1935) 'Essai de l'etude intravitale des cellules nerveuses et des connexions interneuronales dans le systeme nerveux autonome'. *Trab. Lab. Invest. biol. Univ. Madr. 30,* 403-434.

[16] Flaig , J. V. (1947) 'Viscosity changes in axoplasm under stimulation'. *J. Neurophysiol., 10,* 211-221.

[17] Gellhorn, E. (1947) 'Effect of afferent impulses on cortical suppressor areas'. *J. Neurophysiol., 10,* 125-132.

[18] Hecht, Selig. (1955) 'Theory of visual intensity discrimination'. *J. Neurophysiol., 18,* 767-789.

[19] Helmholtz, H. L. Ferdinand von. (1924) *'Treatise of Physiological Optics'.* Translated from the 3d German edn James P. C. Southall, ed. Rochester, New York: The Optical Society of America.

[20] Hovey, H. B. (1928) 'Nature of apparent geotropism of young rats'. *Physiol. Zool. 1,* 550-560.

[21] Hume, David. (1937) *Treatise of Human Nature, Being an Attempt to Introduce the Experimental Method of Reasoning into Moral Subjects.* New York: E. P. Dutton & Co.

[22] Kohler, W. (1929) *Gestalt Psychology.* New York: H. Liveright.

[23] Kraeplin, E. (1904) *Psychiatrie,* 17th edn. Leipzig: J. A. Barth, Vol. 2, p. 482.

[24] Lewy, F. H., and Gammon, G. D. (1940 'Influence of sensory systems on spontaneous activity of cerebral cortex'. *J. Neurophysiol., 3,* 388-395.

[25] Lloyd, D. P. C. (1946) In Howell's *Textbook of Physiology,* 15th edn., John F. Fulton, ed. Philadelphia: W. B. Saunders Co.

[26] Lloyd, D. P. C. (1946) 'Facilitation and inhibition of spinal motoneurons'. *J. Neurophysiol.,* 9, 421-438.

[27] Lloyd, D. P. C. (1946) 'Integrative patterns of excitation and inhibition in two-neuron reflex arcs'. *J. Neurophysiol., 9,* 439-444.

[28] Lorente de No, R. (1943) Chapter in John F. Fulton, *Physiology of the Nervous System.* New York, London and Toronto: Oxford Press.

[29] Maxwell, C. (1867-1868) On Governors. *Proc. Roy. Soc., 16,* 270-283.

[30] McColl, H. (1945) *Servo-Mechanisms.* New York: Van Nostrand.

[31] McCulloch, W. S. (1945) 'A heterarchy of values determined by the topology of nervous nets'. *Bull. math. Biophys., 7,* 89-93.

[32] McCulloch, W. S. *Finality and Form*—Fifteenth James Arthur Lecture, New York Academy of Science, May 2, 1946, at press, Springfield, Ill.: Charles Thomas.

[33] McCulloch, W. S., and Pitts, W. (1947) 'How we know universals'. *Bull. Math. Biophys.,* 9, 127-147.

[34] Moore, A. R. (1945) *Individual in Simpler Form.* Eugene, Ore.: University of Oregan, Chap. 108.

[35] Morison, R., and Bassett, D. L. (1945) 'Electrical activity of thalamus and basal ganglia in decorticate cats'. *J. Neurophysiol.*, *8*, 309-314.

[36] Morison, R., and Dempsey, E. W. (1943) 'Mechanism of thalamo-cortical augmentation and repetition'. *Am. J. Physiol.*, *138*, 297-308.

[37] Penfield, W. (1947) 'Some observations on cerebral cortex of man', *Proc. Roy. Soc., Lond.* ser. B, 134: 329-347.

[38] Penfield, W., and Erickson, T. C. (1941) *Epilepsy and Cerebral Localization.* Springfield, Ill.: Thomas.

[39] Ramon y Cajal, S. (1909-1911) *Histologie du systeme nerveux* Paris: A. Maloine.

[40] Rosenblueth, A., Wiener, N., and Bigelow, J. (1943) 'Behaviour, purpose and teleology'. *Philosophy Sci.* 10, 18-24.

[41] Shurrager, P. S., and Culler, Elmer J. (1940) 'Conditioning in the spinal dog'. *J. exp. Psychol.*, *26*, 133-159.

[42] Shurrager, P. S., and Shurrager, H. C. (1946) 'Rate of learning measured at a single synapse'. *J. exp. Psychol.*, *36*, 347-354.

[43] Sperry, R. W. (1947) 'Cerebral regulation of motor coordination in monkeys following multiple transection of sensorimotor cortex'. *J. Neurophys.*, *10*, 275-294.

[44] Thorndike, E. L. (1911) *Animal Intelligence.* New York: Macmillan, p. 244.

[45] Von Bonin, G., and Bailey, P. (1947) *The Neocortex of Macaca Mulatta.* Urbana, Ill.: University of Illinois Press.

[46] Von Bonin, G., Garol, H. W., and McCulloch, W. S. (1942) 'The functional organization of the occipital lobe'. *Biol. Symp.*, 7.

[47] Whitehead, A. N. (1919) *Principles of Natural Knowledge.* Cambridge, London.

[48] Whitehead, A. N., and Russell, B. (1910-1913) *Principia Mathematica.* Cambridge, London, Vol. 3.

[49] Wiener, N. (1948) *Cybernetics.* New York: John Wiley.

[50] Wiener, N. (1948) *On the Theory of Dense Gases.* Communicated to the Macy Conference on Teleological Mechanisms. Unpublished.

[51] Wittgenstein, Ludwig (1922) *Tractatus Logico-philosophicus.* London: Paul.

THEORY OF THE HUMAN OPERATOR
IN CONTROL SYSTEMS

I. THE OPERATOR AS AN ENGINEERING SYSTEM

Kenneth Craik died in 1945 when he was only 31; despite his youth his influence has been immense—not so much perhaps by way of his direct contributions to experimental psychology but because his approach and entire attitude to research were original. He formulated several problems clearly for the first time: the nature of dark adaptation; the nature of the processes involved in the muscular control of movement; the transformations performed on sensory input by the nervous system. Not only could he state problems appositely, but he was a great experimentalist having the capacity to perform simple but crucial tests using apparatus (often home-made) with great ingenuity.

The two papers which follow are clearly written and most deceptively easy to read—in stark contrast with the great mass of psychological literature. In them Craik outlines a theoretical approach to the 'Human Operator', discussing what sort of system may be involved in the control of the ballistic movements made during the tracking of a target on an oscilloscope screen. He concludes that the human operator acts as an 'intermittent correction servo'; estimating an error, making a single corrective movement; estimating the residual misalignment between target and response and making a further corrective movement—and so on. He suggests that after a movement there is a refractory period during which a second movement cannot be initiated. This intermittency has often been equated with that part of reaction time not taken up by primary sensory processes or nervous conduction. So far, no satisfactory model for either composition of reaction time or for the control of tracking movements has been produced; we have progressed little further than Craik.

(*Continued on p. 130*)

Reproduced from *British Journal Psychology* 38 (1947) 56-61, by courtesy of the British Psychological Society.

THEORY OF THE HUMAN OPERATOR
IN CONTROL SYSTEMS*

I. THE OPERATOR AS AN ENGINEERING SYSTEM

Kenneth J. W. Craik

I. THE HUMAN OPERATOR BEHAVES BASICALLY AS AN INTERMITTENT CORRECTION SERVO

The evidence for this is the periodic or 'wavy' nature of the time-record of tracking errors, showing a spectrum with a predominant frequency of about $0\cdot5$ s with a smaller cluster of frequencies from $0\cdot25$ to 1 s. This periodicity might be attributed to a sensory threshold or 'dead zone', such that misalinements smaller than a certain value evoke no corrective movement; but there is evidence against this. First, the display-magnification is usually such that the misalinements occurring during steady tracking exceed the known threshold (i.e. visual acuity). Secondly, if the rate of the course, or the magnification of the display for a given course, is increased by a certain factor, the periodicity of the corrections is little, if at all, affected; whereas if their periodicity were determined by the time taken for the misalinements to reach a certain 'threshold' value this alteration should shorten the periodicity of the corrections in the same ratio.

Editor's note. The material of this paper and of a second one to be published in the next number of the *Br. J. Psychol.*, was prepared by the late Dr Craik in March and April 1945. It was written substantially in the form in which it now appears, and undoubtedly was intended, not as a final form for publication, but as a preliminary draft for discussion. In view, however, of the very lively and widespread interest that has developed in the problems discussed in these two papers, and of the attractive and original way in which Craik handles them, it seems highly desirable that they should be made public. The present paper deals with the possibility and significance of the construction of mechanical and electrical models for simulating human behaviour in any kind of continued pursuit task. The second paper deals with fundamental physiological and psychological mechanisms which appear to be involved in operations of the same kind.

Most of the work of preparing the articles for publication was done by Miss M. A. Vince.

Consistently with the above principle, we find that the mean error in tracking any given variable-direction course is nearly proportional to the rate of the course, over a wide range of speeds. We should account for this by saying that the faster the course the greater the misalinements that occur in each period between two corrections, in strict proportion.

II. THE INTERMITTENT CORRECTIONS CONSIST OF 'BALLISTIC' MOVEMENTS

For example, they have a predetermined time-pattern and are 'triggered off' as a whole. This behaviour may be contrasted with that of an intermittent correction servo in which, for instance, a follow-up motor is intermittently switched into a circuit in which it runs until it has reduced the misalinement to zero, and this action reduces the input to it to zero so that the motor stops. In the human operator, on the other hand, at a particular instant (i.e. about 0·3 s after the end of the preceding corrective movement), a corrective movement having a predetermined time-course (usually occupying about 0·2 s) is triggered off. The evidence for this is based on studies of reaction time, i.e. on the internal time-lag of the operator, due to the time taken by the sense-organ to respond, for the nerve-impulses to traverse the central nervous system, for the appropriate response to be 'selected' and for the nerve impulses reaching it to traverse the motor nerves. This lag is about 0·2-0·3 s. Thus, if a human operator's limb movement amplitude, or velocity, or acceleration, were determined continuously by the misalinement, continued oscillations of approximately 0·5 s period, and of whatever amplitude they commenced at, would inevitably result. This tendency could be overcome if a misalinement triggered off a ballistic movement of fairly correct amplitude (say ± 10%); the eye would then detect the residual misalinement, which may be composed of two parts—the error in the first ballistic movement, and any movement of the target which has occurred in the meantime. In the absence of the latter (e.g. in aiming at a stationary target) the second corrective movement may again be accurate to ± 10 per cent of its own value so that the misalinement is reduced to 1 per cent of its original value.

Direct evidence bearing on this ballistic behaviour can be obtained in various ways. First it is easy to present a misalinement to an operator, and then screen his eyes just before he makes his corrective movement. A movement accurate to within about 10 per cent will result, if he has previously learned the 'feel' of the control. If he operated like a follower motor, intermittently switched in, such obscuration of the misalinement would be equivalent to cutting the input connections, and the motor would not of course make any further corrective movement.

Physiologists might make a further hypothesis to avoid the 'ballistic' theory. They might say that the visual misalinement, once it has been detected by the eye, becomes translated into a 'limb-movement-misalinement' (i.e. a kinaesthetic misalinement) which acts as the continuous input to the limb until our kinaesthetic (i.e. joint and muscle senses) register 'correct position', or no misalinement. We certainly do possess such a sense, but it is less easy to abolish it experimentally and to see what happens in its absence, than in the case of vision. Patients with *tabes dorsalis* have considerable loss of kinaesthetic sensation, but it is difficult to know how complete this loss is in any particular case.

In any case, the same general argument as before—that a continuous series of oscillations of the initial amplitude would occur if elimination of kinaesthetic misalinement were the sole determinants of movements, owing to the inevitable reaction-time lag—seems to apply. Further, it is possible to show that considerable precision of movement is maintained even when the movements are made so rapidly that they are completed before the kinaesthetic stimulus corresponding to their first approximation to the right position could have 'gone the round' of the central nervous system and controlled the subsequent output-movement; unless indeed reaction-times to kinaesthetic stimuli were vastly shorter than to any other kind. It is, however, possible to show experimentally that kinaesthetic reaction times are very little shorter than auditory, for instance. Thus, we may ask the operator to move a lever against a stiff spring, so as to make a rapid movement to correct a misalinement, and then to return to his starting point. After he has learned the 'feel' of the control (i.e. its gear ratio and spring tension) and is making fairly accurate movements, we suddenly alter the spring tension, so that he over- or under-shoots. A record of this on a fast-moving drum shows that about 0·15 s elapses after he begins his movement, until he is able to begin a readjustment of it, to meet the modification of resistance.

Similarly, it is possible to show that in playing musical instruments, typewriting, sending morse, etc., complicated patterns of movement are executed at a rate which would be impossible if they were continuously governed by the value of the misalinement, with the inevitable reaction-time lag. Apparently they must be individually performed, triggered off ballistically, and the sensory feed-back must take the form of a delayed modification of the amplitude of subsequent movements. Sensory control, in other words, alters the 'internal gear ratio' or amplification of the operator with a time lag and determines whether subsequent corrective movements will be made; it does not govern the amplitude of each individual movement while it is being made. We could make a servo, using existing engineering principles, which would show the features of intermittent, ballistic correction. But this last point—the fact that the sensory misalinement alters not merely the amplitude of the response but the *relation* between the input, or misalinement and the output, or response—introduces a further complication, the nearest approach to which, in engineering, seems to be 'floating plus proportional control'. Even this involves a quantitative alteration of the amplification of the system by the residual misalinement, whereas something more complex still seems to be occurring—a very wide alteration in the functional relationship between input and output.

For instance, if the operator is using a positional control, his successive ballistic corrective movements should be linearly proportional to the misalinements; but if he is using a velocity control they would have to be linearly proportional to the *derivatives* of the misalinements. Roughly speaking, we might call this *qualitative* modification or output on a basis of some response to the difference between instantaneous input and output at the previous instant, or 'qualitative feed-back'.

A further complication is introduced by the operator's ability to 'anticipate' movements of the target, or alterations in misalinement. For instance, with a positional control, the errors in tracking a moving target are usually less than we should predict on the above theory of intermittent ballistic corrections. The operator goes on turning the handle steadily, or even accelerating it; his record, after some practice, becomes much smoother than it was initially; and it he finds that he is still lagging behind the target (as if the above theory is correct he is bound to do) he can put in an extra forward movement. Here we have several processes.

III. THERE ARE SOME COUNTERACTING PROCESSES TENDING TO MAKE CONTROLS SEEM CONTINUOUS

First, there is one akin to momentum, or inertia. If the operator has been turning the handle, in a series of discrete movements, for some seconds, he will tend to convert this into as steady a rotation as he is capable of, and to continue doing so although the misalinement may be zero, i.e. he has zero input to produce this output! (This can be shown experimentally by suddenly stopping the target, when the operator will overshoot for the period of one reaction time, until the serious misalinement which results stops his steady output from continuing.) It is for this reason that in Section I it was stated that the human operator is *basically* an intermittent correction servo; he has in addition this mechanism for going on doing whatever is giving a satisfactory result, or zero misalinement, rather like a heavy flywheel, and having the same valuable smoothing effect.

What are the essential features of this process, and can we conceive any mechanisms which will accomplish it? When the operator continues to turn the handle at the same speed, independently of whether there is any input or not, he is, in human - istic terms, assuming that he is justified in doing so, in order to compensate for his reaction time lag. Since he is always subject to this lag, in attempting to keep up with the present he is always in fact being a prophet and extrapolating from past data! It is really no different from the further kind of anticipation which enables him to extrapolate into the physical future. Now all scientific prediction consists in discovering, in data of the distant past and of the immediate past which we incorrectly call the present, laws or formulae which apply also to the future, so that if we act in accordance with those laws our behaviour will be appropriate to that future when it becomes the present. Thus the essential feature of extrapolation and anticipation is, again in humanistic terms, that the operator should detect the *constants* in what he is doing. Thus, he may move a handwheel in a series of jerks, so that its *position* changes from moment to moment, but he may realize after a few seconds that he is turning it at a *steady rate*, i.e. its *angular velocity* is constant; and having discovered this he may try whether it will not pay him to go on doing so; usually it will. He may, however, find that its rate is changing—the target has angular acceleration. He may, in theory, at any rate, be able to feel this acceleration which he is having to put into the handwheel, and if he happens to find that it is constant or nearly so, and is able to put out a constant acceleration of this value in turning the wheel, again he may achieve better following.

Now let us look at the same thing from a mechanical point of view. There are many devices—such as speedometers and accelerometers—which do the differentiations involved in recording velocities and accelerations; and the problem would be how, for instance, to couple a number of such devices to a telephone selector-switch operating motor, so that if the output of the motor over a few seconds of intermittent corrective action showed a constant reading on the speedometer, or even on the accelerometer, the motor would be caused to go on putting out this speed or acceleration, irrespective of whether there was any input or not, unless or until such behaviour gave rise to a large misalinement. If that happened the extrapolating system would be overridden and intermittent corrections would begin again, until a new value for a constant was found. The solution would seem to be to provide the motor with positive feed-back of such a kind that it continued to go on doing whatever it was doing at the moment—running steadily or accelerating uniformly. Such a system would need considerable smoothing and stabilization, otherwise any slight disturbance, such as a slight acceleration, would very rapidly be cumulative, and the machine would reach its maximum speed ; but if the feed-back were delayed and smoothed, the system could be sufficiently stable and would not 'wander' too badly. This system would of course be combined with negative feed-back of the ordinary kind (namely actuation by the difference between input and output quantities), so that if the positive feed-back led to overestimation of the velocity, or if the target started to decelerate, the motor would overshoot and this would introduce a positional misalinement, which would reverse the direction of mechanical control. This would alter the average value, for the last time-interval, to the positive feed-back system, which would therefore cease from perpetuating this velocity but would, when it had time to steady down, start putting in a new one.

IV. ELECTRICAL MODELS COULD FAIRLY EXACTLY SIMULATE THE HUMAN OPERATOR'S BEHAVIOUR IN TRACKING

In general terms, the extraction of the inputs for the positive feed-back network consists of successive differentiations, while the extrapolations on the basis of them consist of successive integrations. Let us consider in more detail some circuits by which this might be accomplished. Suppose the motor drives a generator across whose output terminals is a capacity in series with a high resistance, constituting a differentiating circuit with a time-lag or averaging effect, owing to the time-constant of the system. Then the generator voltage is proportional to the speed of the motor and the voltage across the resistance of the differentiating system is proportional to its acceleration; if necessary, higher derivatives can be obtained in the same way. The lag in the first differentiation can be obtained by putting a resistance and capacity in series across the generator output and taking the voltage off the capacity. The output voltage from this smoothing system is taken to the input of the amplifier supplying the motor fields, and should cause the motor to continue running at the mean speed at which it has been manually rotated for a sufficient time to cause the voltage delayed and smoothed across the generator to reach a steady value. The speed of running will of course wander slowly in time if the system also has ordinary velodyne negative feed-back for velocity control. If the manually applied speed was an accelerating one, the system will maintain a mean steady speed if it is supplied with one differentiating stage only (i.e. the generator with its delaying system). But if there is a second differentiating system, with a longer time-constant, it will register a manually imposed change of velocity over several periods of operation of the first differentiating system, i.e. an acceleration, and if this is integrated by a resistance-capacity circuit and applied to the amplifier serving the fields, a uniform acceleration will occur.

Of course, it is not necessary for the original speeds to be put in manually; with a velodyne fitted up as a servo auto-following system in which the task of the velodyne is to make a slider keep on the centre of a potentiometer, for instance, which is moved by an external agency, the mechanical control will commence by ordinary positional following, being actuated by the misalinements. Further, though this alone would lead to a lag behind a uniformly moving target, once it has started to run at a constant velocity, the remaining misalinement will still be operative, if arranged to be in series with the positive feed-back voltage, so as to cause the shaft to step on and make up for the lag. This system would have many of the same characteristics as a velodyne with phase advance produced by delayed negative feed-back, i.e. a condenser across the generator output.

It should be possible to make a velodyne simulate the 'intermittent ballistic correction' process considered in principles (I) and (II). Thus, the error-voltage representing the misalinement could be connected periodically by a rotating contact to a condenser which is charged. This condenser would then be switched on to the amplifier input and would result in a 'ballistic' rotation of the output shaft through an angle proportional to the charge on the condenser.

Little is known of possible physiological mechanisms for accomplishing this kind of thing. There is evidence (e.g. from sensory adaptation and accommodation of nerve) of differentiating systems, at least of the first order, which may serve to measure rates of change of stimulation, though our knowledge extends only to stimulus *intensity* and not to more complex stimuli such as misalinements in space. Even here, it is possible to suggest hypothetical spatial differentiating systems which are not physiologically inconceivable. The other aspect—the integration, resulting from positive feed-back—would seem to require 'auto-rhythmic' nervous centres which continue to discharge once they have been forced to do so, and in a way which follows the original forcing stimulus. The beating of the isolated frog's heart and the spontaneous oscillatory potentials in the excised frog's brain and in the intact cortex of man (both Berger rhythms and the abnormal rhythms of epilepsy) are suggestive in this respect, for they are evidence of self-maintaining neural oscillators. Lorente de No's and Ransom's concept of the 'closed neurone circuit' would serve the same purpose. What has to be considered is clearly a form of positive feed-back and the main difficulty in all the cases just mentioned would seem to be that what is required is continuous feed-back of excitation in the form of nerve-impulses following after the neurones have recovered from their refractory phase, whereas slow potential oscillations probably imply discharges of some other kind than trains of nerve impulses.

We should also consider long-lasting changes of stimulus-response relationship (i.e. learning) which, in an electrical model, would probably require to be imitated by some autoselective switching device rather than regarded as time-constants of a resistance-capacity system. Another type of control demands the establishing of complex response-patterns which are 'triggered off' as a whole by the stimulus. Instances are the action of word-habits in typewriting, or of blocks of stimuli in transmitting morse, or of associated movement groups in knitting. These seem to require some 'sequencing' switchgear, of the type used in the Relay Automatic telephone system, and make us think of the physiologists' 'chain reflexes' and of rhythmic reflexes such as walking and breathing.

THEORY OF THE HUMAN OPERATOR
IN CONTROL SYSTEMS

II. MAN AS AN EXAMPLE IN A CONTROL SYSTEM

(Editorial comment continued from p. 120)

Craik's approach was to establish a model for any system being investigated and to test the model rigorously with simple experiments. This is the most difficult, but none the less the only legitimate approach to the notoriously ill-defined problems of psychology. We rarely recognize how much our attitude to psychological research has matured and how much of this change is directly due to Craik.

Further Reading

Bertelson, P. (1965). 'Central Intermittency Twenty Years Later' *Q. Jl. exp. Psychol.* **18**. 153-163

Birmingham, H. P. and Taylor, F. U. (1954). 'A design philosophy for man-machine control systems'. *Proc. Inst. Radio Engrs.* **42**, 12, 1748-1758

Reproduced from *British Journal Psychology* 39 (1947) 142-148 by courtesy of The British Psychological Society.

THEORY OF THE HUMAN OPERATOR
IN CONTROL SYSTEM*

II. MAN AS AN EXAMPLE IN A CONTROL SYSTEM*

K. J. W. Craik

1. STATEMENT OF THE PROBLEM

As an element in a control system a man may be regarded as a chain consisting of the following items:

(1) Sensory devices, which transform a misalinement between sight and target into suitable physiological counterparts, such as patterns of nerve impulses, just as a radar receiver transforms misalinement into an error-voltage.

(2) A computing system which responds to the misalinement-input by giving a neural response calculated, on the basis of previous experience, to be appropriate to reduce the misalinement; this process seems to occur in the cortex of the brain.

(3) An amplifying system—the motor-nerve endings and the muscles—in which a minute amount of energy (the impulses in the motor nerves) controls the liberation of much greater amounts of energy in the muscles, which thus perform mechanical work.

(4) Mechanical linkages (the pivot and lever systems of the limbs) whereby the muscular work produces externally observable effects, such as laying a gun.

Such considerations serve to bridge the gap between the physiological statement of man as an animal giving reflex and learned responses to sensory stimuli, and the engineering statement in terms of the type of mechanism which would be designed to fulfil the same function in a wholly automatic system. The problem is to discover in detail the characteristics of this human chain, such as its sensory resolving-power, its maximum power-output and optimum loads, its frequency-characteristics and time-lags, its amplitude-distortion and whether or not internal cyclic systems enter into it, its flexibility and self-modifying properties,

---

* The first part of this paper appeared in the *Br. J. Psychol.* 38, 56-61.

etc., with a view to showing the various advantages and disadvantages of the human operator as compared with an automatic system.

2. HOW THE PROBLEM MAY BE INVESTIGATED

The following are some of the techniques available for studying the human operator as an element in a control system:

(1) We may set him the task of tracking a real target and record his errors by a cine-camera mounted on a gun, and later plot the actual course of the target from knowledge of the motion of the gun and the laying errors at each moment. His performance with different types of control systems can thus be compared, and *ad hoc* results obtained.

(2) We may set him the task of tracking an artificial target following a predetermined course, e.g. by connecting a differential gear between a shaft whose angular rotation is determined by a cam and represents the course, and his control handle; his task is to keep a pointer fixed to the jockey of the differential on a stationary point, and so to balance the target-shaft motion. The movements of this pointer can be recorded on moving paper and will indicate his errors. This method has many advantages of simplicity and cheapness, and makes it possible to repeat exactly the same target-course on successive occasions. Electrical, hydraulic, and pneumatic equivalents of this mechanical system are of course feasible. We can also set 'unnatural' courses containing step-functions, sudden applications of constant velocity to the target, sinusoidal motion, etc., in order to reveal particular characteristics of the operator's performance.

(3) We can attempt to isolate various steps in the chain composing the operator's response. For instance, by suddenly concealing the display we can break into the cycle of operations and momentarily determine the relation between input and output in the 'straight' system—or at least with the system containing only internal feed-back such as that provided by kinaesthetic sensations (sensory indications of how far a limb has moved). We can determine the limiting resolving power of the sensory system—e.g. visual acuity—the smallest movements a limb can make, and the maximum power of the limbs with their muscles, by magnifying levers. Physiologists have some techniques, such as recording nerve impulses in sensory and motor nerves and in the brain by means of amplifiers and oscillographs and stimulating motor nerves electrically, which are of great physiological

interest, but have as yet, owing to technical difficulties, yielded insufficiently quantitative data to be of great assistance in deriving mathematical laws for the operating characteristics of the human being in a tracking task.

3. THE FIRST TWO STEPS IN THE CHAIN OF RESPONSE

(a) The sense-organ

The eye is almost invariably the most suited for tracking tasks, owing to its high sensitivity and resolving power, its two-dimensional and even three-dimensional appreciation of objects. It can detect misalinements as small as 5 s of arc (as in vernier scales and coincidence range-finders) and can distinguish between neighbouring lines 1 min of arc apart, and it has great power of appreciation of form or pattern, e.g. in distinguishing a tank from the surrounding country, or an enemy aircraft from a friendly one. Only in the distinguishing of a constant frequency from a mixed-frequency background—as in detecting Asdic echoes—or in distinguishing a rhythmic modulation from a steady background of mixed frequencies—as in the beating of a submarine's propellers picked up by a hydrophone—is the ear superior to the eye, and therefore is used for tracking some targets in the early stages of the attack. Occasionally the naked eye may be a limiting factor in tracking, e.g. in visual night air-to-air gunnery, but telescopic sights, radar displays of target position, etc., often exceed this limit.

The first problem we meet in analysing the performance of the visual sense-organ is: 'What is the quantitative relation of the angular misalinement between sight and target to the message sent up the optic nerve as "input" which activates the rest of the human operator-system?' Unfortunately, very little is known about this, except that the receptive surface at the back of the eye contains a mosaic of individual receptors which are individually connected to nerve-fibres, in the region which has the highest resolving power, and that the fineness of this mosaic corresponds roughly with the resolving power of the eye. But how a misalinement between two objects in the retinal image produces something equivalent to the 'error voltage' in a radar set is not known. The problem is further complicated by the fact that the misalinement-image need not fall on the same set of receptors from moment to moment, i.e. tracking can go on in spite of small eye-movements; and of course the particular misalinement to which the operator responds depends on his training and his orders at the time.

Thus, if he is an A.A. gunner he disregards the misalinement between his sights and various enemy tanks or infantry which may be within view, and enemy aircraft apart from that which he is ordered to attack. His training and his orders act as some kind of selector switch in the brain in a way which is completely unknown. It is possible, however, to show that the operator has some quantitative appreciation of the misalinement; he does not act as if he merely had an on-off wiping contact which indicated satisfactory alinement or an unknown degree of misalinement; and he also appreciates the sense or direction of the misalinement. This can be shown by presenting cards with cross-wires and target-dots at various misalinements and asking the operator to state the size of the misalinement when it is compared with a standard misalinement card presented alternately. Ability to do this varies with training and with the time interval between each presented misalinement and the next; but when the cards are presented for 3 s each at intervals of about 1 s, the relation between the physical value of the misalinement and the subjective estimate of it is as in *Figure 1*, which gives the mean values for 10 subjects. It is important in such experiments that the cards should all be presented at the same distance from the eye, otherwise a curious conflict occurs between estimation of the angular misalinement and the actual misalinement in units of length on the card. Thus, if we show an unsophisticated person a card pinned to the wall at 15 ft and ask him to say which of several sizes of cards, which we hold up in turn at a distance of 7 ft, is equal to it, most people will choose some size intermediate between physical equality and and angular equality (i.e. half the linear dimensions) with considerable consistency This type of perception was called 'phenomenal regression' by its discoverer, Thouless[1] and represents an attempt to compromise between judging physical sizes of objects, which is usually the more important, and their angular size which may more occasionally be required.

Thus we may conclude that under suitable conditions the human operator can appreciate the angular size of a visual misalinement to an accuracy of about 10 per cent.

The next point is whether the eye is responding continuously to this misalinement. In a task such as reading, where we wish to observe different parts of the field successively, it can be shown by photography that the eye makes jerks, or 'saccadic movements', having a mean duration of 0·03 to 0·05 s, interspersed with pauses of about 0·3 s, during which alone the retinal image is sufficiently distinct to be appreciated, under normal circumstances. Thus the eye operates intermittently in a task such as reading. In tracking a target, however, similar

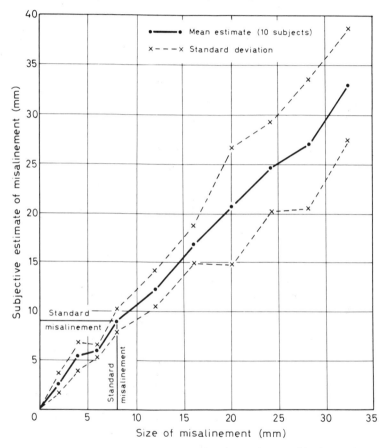

Figure 1. Relation between the physical value and the subjective estimate of the misalinement

photographs show that the eye fixates the aiming-point on the target as steadily as possible, whether the target is stationary or moving, and it can be shown by electrophysiological methods that sensory messages are being sent continuously up the optic nerve. But we shall see later that continuous response by the sense-organ does not imply that the human operator as a whole responds continuously. It may sometimes be desirable, however, to attempt to smooth rapid variations in misalinement, such as may be set up by fading in a radar receiver, or vibration and bumping in tanks and aircraft; partly in order to reduce the blurring of the misalinement in the retinal image, and partly to eliminate those momentary changes in misalinement which are too rapid for the operator to be able to cope with by useful corrective action. This, as we shall see later, is a much slower process, and elimination of rapid oscillations in misalinement may improve the operator's performance by removing his temptation to try to achieve the impossible.

(b) The Computing System

In psychological terminology, the operator learns the feel of the controls and finally makes the control-movements which he judges to be appropriate for reducing the misalinement as much as possible. Viewed objectively, this process consists of a modification of the relation between the input, or optic-nerve message, representing the angular misalinement, and the output, or limb-movement. If no instruction is given, it may take the form of trying and then abandoning various unsuitable methods of moving the controls, for instance with the unsuitable hand, or with an awkward grip, or by action in the wrong direction. There then follows a period of quantitative modification of the ratio of control-movement to the misalinement, i.e. a modification of the 'stiffness' or amplification of the system, and finally there may be an appearance of complicated temporal patterns of control-movements in response to a misalinement, having the object of compensating for the defects of the operator (such as his time-lags) or of the control-gear (such as backlash, striction, and inertia), or overcoming physical limitations such as the time of flight of the projectile (as in aiming off with a shot gun at a flying bird, where no specific rules for aim-off are given). Thus, viewed from the outside, and regarded as a mechanical system such as we should design to operate in the same way, the operator's brain appears as a computing system and amplifier, with variable characteristics and a variable switchgear between its different input and output elements. For this reason it is not possible to give a single equation which will express the brain's behaviour under all circumstances and at all stages of learning. From the practical point of view the most important points to study are probably (a) the 'natural' or untutored response of the human operator to misalinements; (b) the time taken to modify this so as to reach a steady level of performance on any particular kind of control gear; and (c) the limits and defects of his performance when this steady state has been reached. All these points bear on the design of equipment as well as on the training of the operator.

The first and most marked feature of the cerebral process is its time-lag or 'central delay'. The time-lag between the occurrence of a stimulus such as a flash of light, and the most rapid response the operator can make—such as pressing a morse-key—has been measured by psychologists since the last century; for a single visual stimulus it averages about 0·18 s, when a warning signal has been given before. When there is no warning, and when the stimulus takes some time to build up, as with a pointer gradually going out of alinement with a mark, and where the operator has a choice of the direction in which to make his corrective movement and also to some extent of its magnitude, so as to try to match this to the misalinement magnitude, it will be slower—about 0·3 s. Further, in the effort to make a precise and graded response movement he will move the control less rapidly than in pressing a morse-key and the movement itself may take about 0·2 s, making about 0·5 s as the average duration of the whole process. Physiological studies of the response time of the eye and the time taken for impulses to be transmitted up the optic nerve show that these can account for about 0·01 s only, and electrical stimulation of motor nerves and records of the start of the limb-movement which results show that this accounts for another 0·01 s approximately. There are left about 0·28 s to be accounted for in cerebral processes. Practically nothing is known of these, except that the nerve impulses travel through a number of 'synapses' or junctions between nerve-endings which show a delay of about 0·0007 s each. These junctions transmit nerve-impulses in one direction only, from sensory to motor nerves, unlike nerve fibres themselves, which transmit either way. Synapses thus resemble electrical rectifiers; they also store up incoming impulses over periods up to 4 s and liberate a stream of outgoing impulses which may outlast the incoming volley by some 5 s—the after discharge. Thus they alter the time relations of the incoming impulses. They also show variations in ability to transmit impulses—generally a decrease after repeated stimulation—analogous to a rise in electrical resistance in a circuit. These changes are often regarded as the basis for a 'switching' mechanism which would account for learning and modification of behaviour. But all the resistance-changes so far detected are much too short for this, and it is likely that other types of alteration in synaptic resistance are involved in the living animal.

We must therefore ask ourselves whether this delay is more likely to consist of the transmission-time of nerve impulses continuously travelling down an immensely long chain of nerve-fibres and synapses connecting sensory and motor nerves, or of a 'condensed' time-lag occurring in one part of the chain. If the first hypothesis were correct, there would seem to be no reason why a continuous stream of incoming impulses should not evoke a continuous stream of motor ones, just as impulses can follow each other closely in a nerve-fibre or fairly closely even in a synapse. If, on the other hand, the time-lag is caused by the building up of some single 'computing' process which then dis-charges down the motor nerves, we might expect that new sensory impulses entering the brain while this central computing process was going on would either disturb it or be hindered from dis-turbing it by some 'switching' system.

These ideas can be tested to some extent by recording the human response to a series of discrete stimuli—such as flashes of light—presented at various time-intervals, to see whether there is a minimum interval within which successive stimuli cannot be responded to. Such an experiment is analogous to physiological investigations of the 'refractory phase' of a nerve or synapse, as pointed out by Telford[2]. The results of Telford and of the writer suggest a refractory period of about 0·5 s, such that a stimulus presented within this interval after the preceding one is responded to later, or may be missed. If, again, the second stimulus succeeds the first very rapidly—within about 0·05 s—it and the first may by apprehended together and responded to as a single one, as if it had registered before the computing system had started to operate. Stimuli coming in between these two time-intervals are either disregarded, responded to after the first, or cause general disturbance and conflict in the operator,* The result is to set up a response frequency of about 2/s.

It is clear that this limit to response frequency is not deter-mined by the sense-organ as pointed out above, or by the muscle and limb, since a finger can be moved voluntarily as fast as eight times a second, or faster by electrical stimulation; it must apparently be a cerebral limit. Certainly it is possible to make successive movements at a greater pace, as in transmitting morse, typewriting, playing the piano, etc., but in such cases it seems as if either the groups of stimuli composing letters of

---

* For more detailed experimental evidence see the article by M. A. Vince in *Br. J. Psychol.* 38, 149.

morse, for instance, must have been learned previously and that these groups then become single 'stimuli' with which the computing system deals as wholes, or, as in the case of reading music, that the page must be open in front of the player so that he can take in a few notes in succession as one group and respond to them as a unit. He could not play a piece at sight so fast if each note were exposed in turn by a moving shutter.

Thus the operator, in tracking, responds intermittently, at a frequency of about 2/s. Signs of this are often clearly visible in the records, particularly of untrained operators, and it would appear that when it is less evident, in experienced operators, this can be accounted for in terms of the superposition of a smooth movement, thought to be appropriate to the circumstances, and does not indicate any cessation of their basically intermittent response. This conclusion puts the human operator into the class of 'intermittent definite correction servos' apprehending a misalinement, making a single corrective movement, and so proceeding. The next task will be experimentally to analyse the operator's performance, as such, and to see how it is affected by central delay and by distortions or variations in the quantitative relationship between the misalinement at a moment when it is apprehended and the movement which results from it.

REFERENCES

[1] Thouless, R. H. (1931). 'Phenomenal regression to the real object.' *Br. J. Psychol.* 21, 339-359.

[2] Telford, C. W. (1931). 'The refractory phase of voluntary and associated responses.' *J. exp. Psychol.* 14, 1-36.

9

NERVOUS GRADATION OF
MUSCULAR CONTRACTION

In normal posture and locomotion continual adjustments of muscular tension are necessary to counteract the effects of changing load and of fatigue. There are receptors in muscles, the muscle spindles, and tendon organs, which respectively signal the degree of stretch and the tension of muscles. Stretching a muscle, and therefore its spindles which are attached in parallel with the muscle fibres, causes the spindles to discharge, exciting the muscle's motorneurones. A feed-back loop by which information about muscular stretch is used to control contraction therefore exists; it has been the subject of a great deal of research, which has led to a good understanding of the control of posture. This paper describes the work which established the existence of servo control of posture and movement and introduces a method of investigating the properties of the servo-loops concerned.

In characterizing a servo-mechanism one approach is to determine the error, or difference between input and output, by, for example, requiring a human operator to 'track' a spot appearing on an oscilloscope screen. The voltage applied to the C.R.T. plates to deflect the spots can be controlled and compared with that applied by the human operator as he performs his tracking task. The difference between the two is an error signal whose frequency and phase characteristics can be analysed and used to determine the properties of the human operator. While most of the energy in the error signal is concentrated at low frequencies, there is an appreciable peak at about 9 c/s. It is suggested that this represents oscillation in the muscle-length servo loop.

A good account of the neurophysiology involving muscular control is:

Roberts, T. D. M. (1967). *The Neurophysiology of Postural Mechanisms*, Butterworths; London

Reproduced from *Brit. Med. Bull.* 12 (1956) 214-218 by courtesy of The Medical Dept. British Council.

NERVOUS GRADATION OF
MUSCULAR CONTRACTION

H. Hammond, P. A. Merton and G. G. Sutton

When a certain movement is to be made, the motor centres have to bring into action the necessary muscles in the correct sequence and also—and this is the problem with which we are concerned here—to ensure that at each instant the strength of contraction in each muscle is accurately graded to its task, for example, to support the hind-quarters of a cat, or to hold up 5 lb (2·3 kg), or to slew the eyeballs through 40 degrees. The general principle on which the nervous system does this is clear enough. It employs sensory receptors to report how much of the object is accomplished and, in the light of this information, or 'feed-back', it modifies the rate of motor discharge. As a man stretches out his arm to pick up a pencil, information on the position of his fingers is fed back to the brain by his eyes and used to steer them. If he shuts his eyes he will probably miss and be reduced to groping. It might be possible to reach the pencil using only visual feed-back, but in practice several other sorts of information are used, which do not necessarily reach consciousness: for example, positional information from the joints of the arm, touch information from the fingers and information from muscle sense organs—particularly the muscle spindles, whose job it is apparently to monitor the performance of the muscles themselves. If these sources are cut off, the arm becomes almost unmanageable under visual control alone, and the task of getting the pencil, if accomplished, would be very slowly and imperfectly carried out. It is with the muscle spindles that we are largely concerned here, for they seem to have a good deal of the responsibility for seeing that the muscle does just what is asked of it.

1. POSTURAL TYPE OF CONTRACTION:
THE 'LENGTH SERVO'

The muscle spindle lies among the main muscle fibres and share

their attachments, so that they are extended when the muscle
lengthens and relaxed when it shortens (*Figure 1*) (Sherrington[1]).
The reflex connections are such that impulses set up by stretching the spindles excite the muscle's own motoneurones. Thus
extension of the muscle results in an augmented contraction

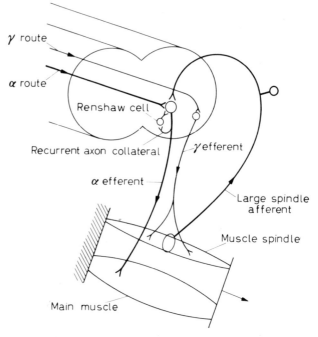

Figure 1. Diagram of the stretch-reflex and related mechanisms

The muscle spindle lies in parallel with the main muscle and its fast conducting afferent is in synaptic connection with the large α motoneurone
supplying the main muscle fibres. A slow conducting γ motor efferent (thin
line) supplies the contractile poles of the spindle and thus can alter the bias
on the spindle sensory ending. The muscle can be made to contract either by
impulses from higher centres exciting the α motoneurone direct (the α route)
or by impulses in the α efferents (the γ route) which activate the muscle
indirectly via the stretch-reflex arc (the 'follow-up' servo). A subsidiary
feed-back via the recurrent axon collateral and an inhibitory Renshaw inter-
neurone may be concerned in stabilizing the response of the α motoneurone to
its excitatory input.

which tends to resist the extension (*Figure 1*). This mechanism
is the 'stretch-reflex' (Liddell and Sherrington[2]) and we see it
has the properties of a 'length servo', i.e., a self-regulating
closed-loop mechanism using negative feed-back from the
spindles to maintain a constant muscle length.

The advantages of a length servo over a simple straight-through system (in which posture is maintained by a steady stream of motor impulses to the muscle without sensory feedback) is that it automatically compensates for changes in load or for fatigue (*see* Eldred, Granit and Merton[3]) and the paper by Merton in this number of the *Br. Med. Bulletin*, 12 p. 219.

Evidence that during a postural type of contraction the muscle really is under this kind of servo-control first came from investigation of the 'silent period'. If a muscle is made to twitch (e.g., by an electric shock) during a steady contraction, the

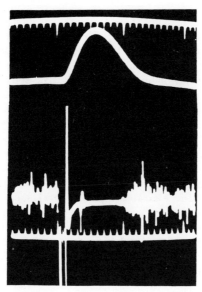

Figure 2. The silent period in the adductory pollicis muscle of the human hand

The subject makes a steady voluntary contraction of 1 kg, and the associated action potentials (the electromyogram) are recorded. When the muscle is excited by stimulating the motor nerve with an electric shock, the electromyogram is silent during the superimposed twitch. This is because impulses from the muscle spindles cease during the twitch (*see Figure 1*) and thus excitation is withdrawn from the α motoneurones (Merton[6])

The record consists of ten traces superimposed, to demonstrate the great regularity of the phenomena.

Twitch tension: roughly 1 kg

Action potentials recorded with a needle electrode

Time markers (top and bottom): 10 ms and 100 ms

electromyogram is silent during the superimposed twitch. The interpretation is that muscle shortening occurs during the twitch, which slows the discharge from the spindles (as demonstrated by Matthews[4]) and thus withdraws excitation from the moto-neurones. A steady voluntary effort in the human subject also seems to employ the length servo, for a silent period is easily demonstrated during voluntary effort (*Figure 2*), and analysis shows that it has the mechanism of a length servo.

Conversely, stretching a muscle during voluntary effort calls forth an opposing contraction (Hammond[5]) which from its brief and constant latency, is clearly a stretch reflex. In these experiments the subject pulled steadily on a wristlet, using his elbow flexors. Without warning the wristlet was pulled away at a constant velocity and the tension in the connecting tape recorded. The subject was instructed to resist the impending pull. Typical responses are shown in Plate IA. The onset of extension is marked by a sharp rise in tape tension as the suddenly applied velocity accelerates the effective mass of the limb. Subsequently there is a transient oscillation in tension, due to a speed fluctuation of the driving motor in response to the sudden loading. During this period the mean level of tension is increased above initial tension by the passive force needed to extend the muscles at a constant rate. After a latency varying from subject to subject, but lying between 60 and 80 ms, the tension in the tape exhibits an abrupt rise. Electromyograms from surface electrodes over biceps show an outburst of activity starting slightly before the tension rise. Earlier in the electromyogram record, some 15-20 ms after the start of the pull, is a diphasic wave representing the tendon jerk. (As observed in cats by Denny-Brown[7], the tendon jerk elicited when a muscle is already contracting does not cause a detectable mechanical twitch, probably because the diphasic wave only involves fibres which were about to discharge anyway, so that the mean rate of discharge is little affected and no significant change in tension occurs.)

An experiment to determine the reaction time to a sharp tap on the wristlet yields the records in Plate I, B. Here the subject is instructed to pull the tape as hard as possible when his wrist is tapped. The diphasic wave in the electromyogram occurs as before, but the delay between tap and rise in tape tension is at once longer and less consistent than was the reaction to an extension. Forcible extension of the flexor muscles thus produces an almost maximal response in roughly half the time taken to react to a tap. The consistency and short latency of the muscular response to an extension lead to the conclusion that here a purely spinal reflex is at work, whereas the response to a tap involves pathways in the brain. It is clear that the shorter latency in response to extension, compared with the latency to a tap, stems from the sustained stretching of the muscle and that the response is that of a stretch reflex or, in other words, a positional servo-mechanism which resists externally imposed postural changes.

It was observed that the rate of rise of tension during the stretch reflex was the same as in a maximal voluntary pull against a stop and was therefore probably as fast as the muscle could achieve (Merton[8]).

2. MUSCLE SHORTENING AND LENGTHENING: THE 'FOLLOW-UP LENGTH SERVO'

The stretch reflex so far described is excellent for holding a fixed posture, but such inflexible behaviour would soon be embarrassing to an animal. The way round this difficulty was first seen by Rossi[9] who remembered that the poles of muscle spindles are contractile and receive a motor supply (called, from its fibre size group, the γ efferents, the main muscle being supplied with larger α efferents). The sensory portion of the spindle lies between the contractile poles, hence shortening of the poles will have the effect of 'biasing' the sensory ending so that the same rate of discharge will occur at a shorter muscle length (Matthews[4]; Leksell[10]; Kuffler and Hunt[11]). Consider a muscle in postural contraction at a certain length. If motor discharge in the γ efferents causes shortening of the spindle poles, this will stretch the sensory portion, so that the stretch reflex will be activated just as if the muscle itself had been stretched. The muscle will shorten reflexly until the increased rate of spindle discharge is offset, and that will be (to a first approximation) when the muscle has shortened to the same extent as the spindles. Hence the complete stretch reflex mechanism, including the γ efferents, constitutes a 'follow-up servo', the muscle length tending to follow changes in spindle length. The spindle endings themselves do not in fact record length, but the differences in length between muscle and spindle; they are misalinement detectors.

To establish the theory, Eldred et al.[3] observed the contraction of the gastrocnemius-soleus muscle of a cat and at the same time listened to the discharge in one of its muscle spindles. To do this it was necessary to cut only a small strand of a single dorsal root, so that the innervation of the muscle was not significantly reduced. As was known from earlier work, various manoeuvres—such as twisting the ear—which caused contraction of the muscle also resulted in great acceleration of the spindle discharge (*Figure 3A*). This implies a rapid γ discharge to the

Plate I. Responses to sudden stretching of the forearm flexor muscles in the human subject

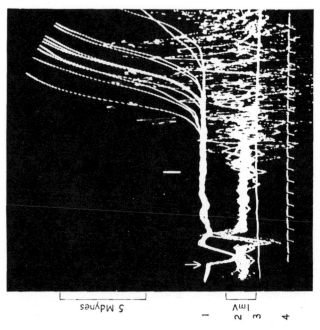

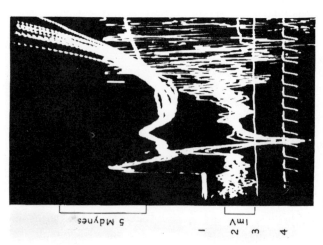

A: subject instructed to resist impending pull. Ten superimposed records of pulls at 40 cm/s at the wrist.

B: ten responses to taps on the front of the wrist delivered at the instant marked with an arrow. Subject instructed to pull hard as soon as he feels the tap. Electrode connections reversed compared with A.

146

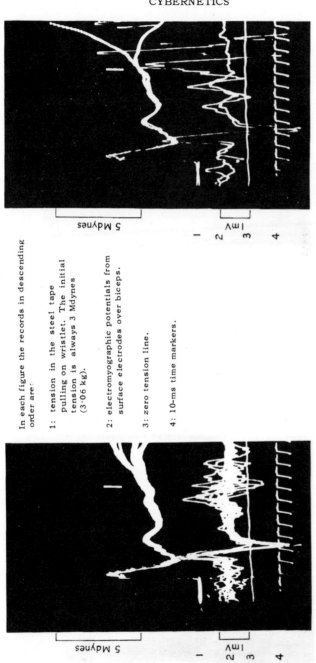

In each figure the records in descending order are:

1: tension in the steel tape pulling on wristlet. The initial tension is always 3 Mdynes (3·06 kg).

2: electromyographic potentials from surface electrodes over biceps.

3: zero tension line.

4: 10-ms time markers.

C: subject instructed to let go when pulled. Conditions otherwise as in A.

D: comparison of one 'resist' and one 'let-go' response. Conditions as in A and C.

On each figure a vertical white bar above the tension trace marks the average time of the abrupt tension rise in the stretch reflex

147

spindles, making them contract faster than the main muscle, for a spindle 'in parallel' with the main muscle stops discharging when shortening occurs. Clearly this behaviour is just what is to be expected if the γ efferents are operating a follow-up servo. Finally the servo-loop was opened by severing all the relevant dorsal roots. Afterwards no contractions were obtained, but acceleration of the spindle was observed just as before (*Figure 3B*).

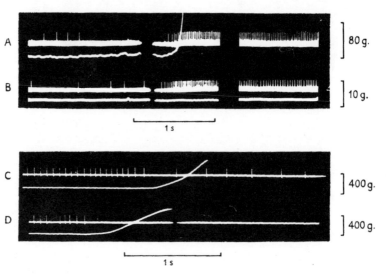

Figure 3. Muscular contractions initiated via the follow-up servo and via the direct route

Adapted from Figure 10 of Eldred, Granit and Merton and Figure 9 of Granit Holmgren and Merton by permission of Journal of Physiology

In these experiments on decerebrate cats, contraction of the ankle extensors is induced reflexly by moving the head up and down. Muscle tension is recorded (continuous line) together with the spike discharge from a single muscle spindle. The interruption of the traces in A and B signals the time of head movement, but was late in *Figure D*.

A, B: decerebration by intercollicular section
  A: muscle contraction accompanied by great acceleration of spindle discharge, indicating that the follow-up servo (γ route) is activated
  B: this is confirmed after cutting the dorsal roots, when spindle acceleration occurs as before, but there is no contraction visible even with the greater sensitivity of the myograph
C, D: another cat decerebrated by tying the basilar artery. This also kills the anterior lobe of the cerebellum and favours the direct (α route) of excitation
  C: contraction is not accompanied by spindle acceleration
  D: after cutting dorsal roots contraction occurs as before. The α motoneurones are no longer dependent on spindle drive

3. THE DIRECT PATHWAY

While the existence of a follow-up length servo seems to be as well established as most concepts in this field, it is perfectly clear from many experiments on de-afferented limbs (in which the servo is necessarily out of action) that movements can often occur without it. Indeed the delay entailed in conduction time down the small γ efferents and up the spindle afferents (which could amount to as much as 50 ms in the case of the hand) suggests that sudden movements might take a short cut direct to the muscle through the α efferents. Granit, Holmgren and Merton[12] showed that, by cooling the anterior lobe of the cerebellum (or by other interference in that region), it was possible to re-route reflex contractions so that exactly the same reflex (e.g., a contraction of gastrocnemius from pinching the ear) which initially used the follow-up servo now used the direct route. This was evidenced by a reversal of spindle behaviour; the sample spindle which was under observation now slowed during contraction instead of greatly accelerating (*Figure 3C*). Cutting the dorsal roots had no influence on the reflex contraction, whereas previously it had abolished it (*Figure 3D*). These results show that the γ efferents are no longer activated during the reflex and that the servo-loop is not in use, as it can be opened with impunity. However, γ paralysis is not the only effect, for, if it were, the reflex contraction should be impossible to elicit (as it was in the previous instance after de-afferentation). As the reflex excitability did not change consistently, the excitation withdrawn from the γ motoneurones must be redirected to the α route. The tentative conclusion is that the cerebellum (and associated structures) determine whether the length servo or the straight-through mechanism is to be used in contraction. This fits in with and extends the current notion that the cerebral cortex decides *what* is to be done and the cerebellum has the job of arranging *how* it is to be carried out.

On the human subject it has also been found that a given contraction can be carried out in two ways, depending on the subject's intentions (Hammond[13]). The responses in Plate I, A were obtained with the subject attempting to resist a pull; those in Plate I, C show the effect of instructing the subject to let go when pulled. It is apparent that prior instruction exerts a profound influence on the vigour of a response. The tension responses to the instruction 'let go' are very much more erratic in shape than are the responses to the instruction 'resist'; this is in accord with a reflex mechanism which favours maintenance of posture as opposed to acquiescence in postural changes, but which the subject can be trained to overcome. The point of divergence of the tension rise as between 'resist' and 'let-go' instructions occurs where the 'resist' responses first begin to rise. Since the abrupt tension rise in a 'resist' response is due to a stretch reflex, it follows from the evidence in Plate I, A and C that the subject can interfere with the stretch reflex to the extent of suppressing it altogether. If the executive organ has control of reflex activity in this way, then its scope is clearly extended to situations where the presence of a stretch reflex might be a disadvantage to the performance of a controlled movement.

1. THE RENSHAW FEED-BACK LOOP: THE TENDON JERK

There is now evidence that, even when the α route is in use, motoneurone discharge is subject to servo-control. Renshaw[14] found that, when an antidromic volley was backfired into part of a motoneurone pool, the other unexcited motoneurones were inhibited. The mechanism has been elegantly analysed by Eccles, Fatt and Koketsu[15] using Eccles' celebrated technique of recording from inside the motoneurone with a micro-electrode. Branches ('collaterals') from the motor axons run back into the ventral horn and excite internuncial 'Renshaw' cells which in turn inhibit the neighbouring motoneurones. There are various complicating factors that make it far from clear whether it is fair to regard this as merely a negative feed-back circuit stabilizing the response of the motoneurone pool in the same way that voltage feed-back stabilizes an amplifier, but the experiments of Holmgren and Merton[16], who used a natural reflex discharge and studied the effect of varying the reflex tension and the size of the antidromic volley, showed that it could behave as if it were such. If this function is definitely established it will be of great interest, as recurrent collaterals are found throughout the nervous system.

Of particular relevance to experiments in muscle excitation is the existence of an appreciable delay (some 2-5 ms) around the Renshaw loop. This implies that a sufficiently brief input (e.g. a synchronous nerve volley) will discover the sensitivity or 'gain' of the loop as it would be without feed-back, because it will go through before feed-back can act. A most important question to be answered is whether the tendon jerk (or the monosynaptic reflex, which is equivalent) is to be identified with this 'unfed-back' response, and it must be admitted that many difficulties might be resolved if it were. Granit[17] has well described the inconsistencies that arise when the state of excitability of motoneurones is tested by eliciting a reflex contraction, or by a tendon jerk or monosynaptic reflex. This discrepancy is also obvious clinically, although little insisted on; for instance although resistance to passive movement (an index of the stretch reflex) and increase of tendon jerks go together in spastic states, the reverse is so in cerebellar disease where we may find absolutely flaccid limbs with brisk jerks.

A clear distinction between jerk and stretch reflex was also apparent in the experiments on the human stretch reflex with 'let-go' and 'resist' instructions described above. It was always observed that the diphasic wave was unaffected by the instructions even when the changes in stretch reflex were most complete (Plate I, D).

The human evidence thus certainly shows that the tendon jerk and the stretch reflex must be considered separately. The reason for the importance of the tendon jerk in medicine may be that it reveals changes in the motoneurones which in ordinary activity are compensated by feed-back. If this hypothesis is correct, the status of the tendon jerk is reduced to an accidental overload condition of a nervous pathway.

5. ACCURACY OF MUSCULAR CONTROL

Another approach which yields information about the mechanism of muscular control is to measure the accuracy with which a muscular effort can be made. The equipment used was designed for work on manual tracking (Sutton[18]), but in the experiments to be discussed the operator had merely to press on a joystick (using his wrist muscles) with a fixed force of, say, 5 lb (2·3 kg) A voltage proportional to this force was generated and used to deflect the spot on a cathode-ray tube facing the subject, whose task it was to hold the spot at the 5-lb mark. The error could either be recorded on film for analysis (*Figure 4*) or squared and integrated by electrical means at once.

When the demanded force was altered in successive experiments the relative accuracy (measured as the coefficient of variation) remained the same. This means that the sensory feed-back mechanisms used to carry out the task treat as of equal significance errors that are a constant fraction of the force exerted, as in Weber's law. Merton[6] showed that the change in tension needed to produce a silent period was similarly related to the initial force.

Frequency analysis of the error record (*Figure 5*) shows that the greater part of the error is at low frequencies with a small peak around 9 c/s, as noted by previous workers (e.g. Travis and Hunter[19]; Sollenberger[20]; Redfearn and Halliday, unpublished observation). Interest has centred on the 9 c/s peak because it is felt that, whereas the large activity at lower frequencies represents the total of the tracking performance of all the feed-back

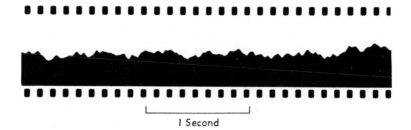

Figure 4. Record of tremor in a voluntary task

The subject is attempting to exert a pressure of 5 lb (2·3 kg) on a joystick. His inaccuracies, which are shown in the record, are only a few per cent of the total force. The 9 c/s component of tremor (shown more clearly in the harmonic analysis in *Figure 5*) can be distinguished here.

loops involved, the 9 c/s peak may be due to slight oscillation of one particular loop, e.g. the length servo-loop. If this were so, any interference with the loop might be expected to modify the 9 c/s peak. Preliminary experiments have shown that moderate cooling of the forearm (in water at 15° C for 20 min) profoundly decreases 9 c/s tremor, which suggests that the 9 c/s peak is more likely to be due to oscillation in the muscle servo than, for instance, to modulation of descending activity by the α rhythm.

With this experiment in mind it was interesting to find that the 9 c/s peak could also be reduced by withdrawing visual information during the task. At the end of a 30 s run with normal visual information in the above experiment, the subject was asked to shut his eyes and continue the task with proprioceptive information only, for an equal time. The great reduction in the 9 c/s peak is shown in *Figure 5* and the effect has been fully substantiated in numerous experiments on four subjects. (Similar results were obtained when the subject kept his eyes open but the cathode-ray tube spot was extinguished.)

In themselves the eyes-closed experiments might suggest that 9 c/s tremor was a modulation imposed by the highest centres but, if the view that 9 c/s tremor is a resonance in the muscle servo-loop is to be retained, it must be concluded that withdrawing visual feed-back reduces 9 c/s tremor by altering the parameters of the muscle servo-loop; not that such a possibility causes surprise, in view of what is now known of the α and γ modes of contraction.

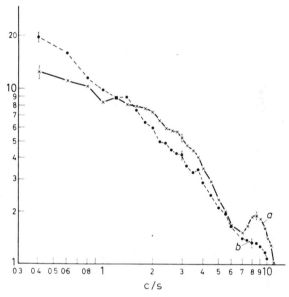

c/s

Figure 5. Frequency analysis of the errors in making a voluntary effort

Abscissae: frequency c/s on a logarithmic scale
Ordinates: relative amount of error at each frequency expressed as error, in
        arbitrary force units, per unit band width on a logarithmic scale
The subjects were attempting to maintain a constant force of 5 lb (2·3 kg).
Each point is the mean of 16 experiments. The vertical bars give the standard
deviation of the 16 observations about the mean.
With the eyes open (curve (a)) there is a peak at 8·5 c/s which is absent when
the eyes are closed (curve (b))

6. MUSCLE SPINDLES IN EYE MUSCLES

The recent discovery of typical muscle spindles in the eye muscles of man and some other animals (see Cooper, Daniel and Whitteridge[21]) is not least important in furnishing proof that

153

muscle spindles are not concerned in position sense but are, in Granit's phrase, the private measuring instruments of the muscle's servo-mechanisms. As shown by Helmholtz[22] there is no position sense in the eye muscles; passive movements of the eyeball are unappreciated, and the resulting shifts of the image on the retina are interpreted precisely as if the eye had stood still and the external world had moved. The eyes are directed by judgement (based on past visual experience) of the amount of effort needed to move them. If the relationship between effort and resultant movement is grossly disturbed, for example by curare, the position of external objects is misjudged and eye movements are accompanied by apparent movement of the external world, as one of us (P.M.) has recently confirmed.

Thus none of the extremely numerous proprioceptive endings in the eye muscles have access to consciousness, and experiments recently begun are confirming the old view that the same is true of the tongue and limb muscles. Sherrington[23] insisted that the sensory endings in eye muscles subserved position sense, but that must have been chiefly because he could not think of anything else for them to do. The function of the muscle spindles, as at present comprehended, is to ensure by feed-back action that the contraction is accurately related to the effort made.

In cerebellar ataxy the patient is absolutely unable to make his muscles do as he wishes. Granit et al.[12] suggest that this is because the muscle spindles are paralysed, depriving him of his muscle servo. No loss of sensation whatever occurs in cerebellar disease except that the judgment of weights held in the hand is impaired (Holmes[24]). This would be unintelligible if the spindles were position indicators, but if they serve to make accurate the sense of effort, it is easy to see why weights should be misjudged. Thus, although messages from muscle spindles do not reach consciousness, the spindle system may be involved in false sensory judgements—for example, apparent movement of objects in the visual field after curare, or erroneous estimation of weights in cerebellar disease.

REFERENCES

[1] Sherrington, C. S. (1900). In: Schafer, E. A., ed. *Textbook of Physiology*, p. 1010. Pentland, Edinburgh and London.

[2] Liddell, E. G. T. and Sherrington, C. S. (1924). *Proc. Roy. Soc.* B, *96*, 212.

[3] Eldred, E., Granit, R. and Merton, P. A. (1953). *J. Physiol., Lond.* *122*, 498.

[4] Matthews, B. H. C. (1933). *J. Physiol., Lond. 78*, 1.

[5] Hammond, P. H. (1954). *J. Physiol., Lond. 127*, 23 P.

[6] Merton, P. A. (1951). *J. Physiol., Lond. 114*, 183.

[7] Denny-Brown, D. (1928). *Proc. Roy. Soc., Lond. 103*, 321.

[8] Merton, P. A. (1954). *J. Physiol., Lond. 123*, 553 .

[9] Rossi, G. (1927). *Arch. Fisiol. 25*, 146.

[10] Leksell, L. (1945). *Acta. physiol. scand. 10*, Suppl. 31.

[11] Kuffler, S. W. and Hunt, C. C. (1952). *Res. Publs. Ass. nerv. ment. Dis. 30*, 24.

[12] Granit, R., Holmgren, B. and Merton, P. A. (1955). *J. Physiol., Lond. 130*, 213.

[13] Hammond, P. H. (1956). *J. Physiol., Lond. 132*, 17 P.

[14] Renshaw, B. (1941). *J. Neurophysiol. 4*, 167.

[15] Eccles, J. C., Fatt, P. and Koketsu, K. (1954). *J. Physiol., Lond. 126*, 524.

[16] Holmgren, B. and Merton, P. A. (1954). *J. Physiol., Lond. 123* 47 P.

[17] Granit, R. (1955). *Receptors and Sensory Perception*, pp. 219, 236, 273. Yale University Press, New Haven; Cumberlege, London.

[18] Sutton, G. G. (1956). *J. Physiol., Lond. 132*, 7 P.

[19] Travis, L. E. and Hunter, T. A. (1931). *J. gen. Psychol. 5*, 255.

[20] Sollenberger, R. T. (1937). *J. exp. Psychol. 21*, 579.

[21] Cooper, S., Daniel, P. M. and Whitteridge, D. (1955). *Brain, 78*, 564.

[22] Helmholtz, H. (1867). *Handbuch der physiologischen Optik*, sect. 29, p. ˉ598. Voss, Leipzig. (English translation by J. P. C. Southall, 1925. *Helmholtz's treatise on physiological optics*, vol. III, sect. 29, p. 242. Optical Society of America).

[23] Sherrington, C. S. (1918). *Brain, 41*, 332.

[24] Holmes, G. (1917). *Brain, 40*, 461.

THE ENGINEERING APPROACH TO THE PROBLEM OF BIOLOGICAL INTEGRATION

Unlike an artificial electronic computer the cerebral cortex is surprisingly immune to local damage; this property is undoubtedly due to the great redundancy both of elements and interconnections built into the cortex. It is clear that the connections between the vast numbers of cortical neurones cannot be precisely determined genetically and indeed it is likely that there is considerable scope for error in both genetically determined and 'learned' connections. If there are errors in connection, then it is necessary, if the cortex is to compute with a low error rate, for there to be some redundancy. Cowan has shown that the 'Noisy channel coding theorem' of Shannon can be extended to cover computation with unreliable elements; the theorem can therefore be used to prove that reliable machines can be built from unreliable components. The redundancy entailed in producing this reliability leads to a diffuseness of structure in which many functions are computed by the same element and in which many elements contribute to the computation of the same function. This leads to a delocalization of function as well as to some degree of 'equipotentiality' (as Lashley pointed out for the cerebral cortex as a result of his experiments to determine the decrement in behaviour produced by cortical lesions in the rat). It is interesting that Lashley's results which at first seemed hard to reconcile with the properties of any easily visualized physical system, follow naturally from the necessary properties of the cerebral cortex.

Further reading
Lashley K. S. (1929). *Brain Mechanisms and Intelligence*, University of Chicago Press, but now reprinted.
Wiener. N. (1948). *Cybernetics*, reprinted by Wiley.

Reproduced from J. D. Cowan. *Symposium on Cybernetics of the Nervous System.* 1962 Elsevier; Amsterdam.

THE ENGINEERING APPROACH TO THE PROBLEM OF BIOLOGICAL INTEGRATION

J. D. Cowan

It has been estimated that something like 4×10^{19} bits/s are processed in macromolecular synthesis, and that macromolecular systems containing *blueprints* for future development are produced by cells and bacteria, at error-rates of the order of 10^{-10} per bit (Quastler[1]). It has also been estimated that the actual error-rate per molecule involved in macromolecular synthesis is is of the order of 10^{-2} (Pauling[2]). Assuming a requirement of 5 bits per molecule, we are led to the conclusion that the *system* error-rate is much lower than the *component* error-rate. This suggests that within the macromolecular system there exist various integrating mechanisms that effectively control malfunctions in one way or another, to produce a reliable system.

The vertebrate central nervous system apparently displays similar integrative features. It has been estimated that some 2×10^{13} bits/s are involved in signal processing by approximately 10^{10} neurones. Each of these neurones receives many inputs, and apparently computes some fairly specific function. It is not known how reliable this computation is, but estimates of the intrinsic fluctuations of peripheral nerve fibres (Verveen[3]) suggest an error-rate of 10^{-2} per bit as a not unreasonable figure. In addition, neurone interconnectivity does not appear to be rigidly controlled, and a certain amount of local *randomness* appears to exist in many cortical areas. Thus large parts of the cortex appear to comprise heterogeneous populations of units whose interconnections are not completely determined, and whose function is not completely error-free. The average error-rate of the cortex as a whole is something that is difficult in the extreme to determine, and no estimates of this number are known to the author. There is, however, some evidence that the cortex as a whole exhibits an error-rate considerably less than 10^{-2} per bit (McCulloch[4]). This suggests that there exist, at this level of organization, mechanisms for the integration of nervous activity, that produce reliable behaviour, i.e. both molecular and neural populations apparently contain mechanisms that ensure stability of output despite various perturbations of the system (Weiss[5,6]).

Various models have been devised to account for these properties. In dealing with macromolecular synthesis, Matthysse[7] has shown how the mere complexity and the number of possible chemical pathways provide an essential *shock-absorber* against perturbations of the macromolecular synthesizing system. Similarly, Ashby[8] in a consideration of the properties of randomly assembled nerve nets or automata found that stability was exhibited by such nets provided they possessed a moderate degree of multiplicity of interconnections, and provided the components were not all alike. A somewhat similar approach was made by Cragg and Temperley[9] who stressed the importance of having a non-specific organization. They constructed a co-operative model for neural activity based on a suggested analogy between the cortex and the ferromagnet, and noted that the non-specificity of organization would result in a degree of immunity to lesions and malfunctions not present if specific circuits were used. This point was also made by Beurle[10] who studied the spread of activity, essentially co-operative in nature in a mass of randomly connected units which could regenerate pulses. Each unit served as a common storage point for information concerning a large number of occurrences, and information about a particular event was stored in a large number of places. Beurle commented that this *multiple diversity effect* would readily account for the equipotentiality of limited regions of the cortex (Lashley[11]). This organization also appears to be responsible for the immunity to a certain amount of damage of perceptrons (Block[12]).

In a certain sense, these integrative properties are obtained rather fortuitously. The models themselves represent the outcome of a process wherein much of the known structure is axiomatized of the biological system in question, and the resultant behaviour is then studied. This approach is essentially that of the natural philosopher or scientist. There exists a complementary approach, however, that of the engineer, which consists in axiomatizing or specifying behaviour, and then attempting to design or construct structures that will realize it (Uttley[13]). The engineering approach to problems of biological integration was initiated by McCulloch and Pitts[14], Wiener[15] and Von Neumann[16] in their studies of cybernetics and automata theory. The problem of biological integration, now formulated as that of constructing or designing reliable automata from components of low reliability, has been studied extensively in recent years. One of the earliest results was obtained by Von Neumann[17] who showed how *redundant* automata (i.e. automata comprising many more components and connections than would be absolutely required to execute given processes) could be used to perform given computations (logical decision-making and coding) at much lower error-rates than those of the given components. This property resulted from the use, in place of single components, of redundant aggregates of like components that operated on a repeated signal, on a *majority logic* principle ('what I tell you three times is true') to produce a correct output. The organization of these aggregates was very much more specific than the non-specific systems previously discussed, and consisted of many repeated circuits with some small degree of interaction between them. Only a certain amount of local randomness of interconnections was permitted. The reliability of such automata increased with their redundancy, but at a rather slow rate, namely redundancies of the order of $10,000:1$ were required, given components having error-rates of 5×10^{-3} per bit, to obtain overall error-rates of 10^{-10} per bit. It was pointed out by Allanson[18] that a more plausible model for integrative structures in the central nervous system would be one wherein much more complex components existed, having many inputs rather than the few inputs of Von Neumann's elementary components. It was shown how such complex components might themselves control synaptic errors, in a more efficient manner than redundant aggregates of simple components, simply by using replicated synapses together with the majority principle. Muroga[19] and Verbeek[20] carried this work a step further by combining both these techniques. That is, redundant aggregates of complex components were designed to control both synaptic and computational errors. The redundancies required were much smaller than $10,000:1$. Specific models for neural organizations designed to execute given processes such as classification (Uttley[21]) or mnemonization (Roy[22]) have utilized either the majority logic principle or else simple circuit replication.

It will be seen that essentially two different types exist of models for biological integration. The first type, based on an axiomatization of the known structures of biological systems, results in models having a heterogeneous and diffuse structure, while the second type, based on an axiomatization of the required behaviour of integrated systems, results in models having a more homogeneous and well-localized structure. The fact that these latter structures bear little resemblance to biological structures is not surprising. Axiomatization of behaviour does not result in a unique structure; in principle many different structures could realize the desired behaviour. It is also to be noted that the specific structures appear to be less efficient (require more components), in achieving given error-rates than do the non-specific structures. It is clearly of interest to consider other ways of constructing reliable automata that are both more efficient and more diffuse in their organization.

A recent development in information theory (Cowan and Winograd[23]) in fact contains in essence methods for realizing such automata. It has been shown that the noisy channel coding theorem (Shannon[24]) may be extended to deal with *computation* in the presence of noise, in addition to communication, provided certain assumptions are made. The results of this are that reliable automata may be constructed from components of low reliability, in such a way that much smaller redundancies are required than those in earlier constructions.

It has been demonstrated that errors of interconnection may also be combated, so that for large enough automata, both component behaviour and interconnectivity may be to some extent random, yet such automata may still function with very low error-rates.

The type of organization required in such automata is of a diffuse nature: each component computes some composite function of many of the functions that have to be executed by the automaton, and any one of these requisite functions is executed by many different components. The resultant diffuseness, or multiple diversity of structure and function, is associated with very low error-rates in the overall functioning of the automaton. Such automata are in fact (a) heterogeneous and non-specific in functional organization, (b) not completely determinate in structural organization, and (c) highly efficient.

There thus appears to be a link between the scientific and the engineering models of integrated biological systems. Uttley[21] et seq. has in fact shown how specific replicated circuits might arise by chance, in a non-specific or randomly connected network, thus providing the basis for an indirect connection between the two sorts of models. If a direct link between such classes of models can be shown to exist, then *a fortiori*, the engineering approach to problems of biological integration becomes more meaningful. It is our intention to demonstrate this link, and to apply the statistical theory of information processing in automata to the construction of concrete models of such biological systems as the cerebral cortex, and certain macromolecular complexes.

ACKNOWLEDGEMENT

This work was supported in part by the U.S. Army Signal Corps, the Air Force Office of Scientific Research, the Office of Naval Research, and in part by the U.S. Air Force (Aeronautical Systems Division) under contract AF33(616)-8783.

SUMMARY

The necessity to use many-valued logics and/or information theory in the case in which noisy units and noisy connections are *given*, is discussed. We demonstrated that redundant computers exhibiting arbitrarily low frequencies of error (apart from errors in the final outputs) may be constructed so that they are not completely redundant, but process a finite fraction of information. This depends critically upon the error behaviour of components as a function of complexity. If component errors increase with complexity, this reliability may be obtained only by decreasing the fraction of information processed in the computer. However, it is possible to maximize this fraction for given components and codes. Another important result is that such a computer need not be precisely connected and, in fact, a certain bounded fraction of errors in connection may be tolerated. The application of these results to the construction of mathematical models of cortical structure is considered.

REFERENCES

[1] Quastler, H. (1957). 'The complexity of biological computers'. *IRE Trans. EC-6, 3,* 191-194.

[2] Pauling, L. (1960). *Errors in Protein Synthesis.* Festschrift für A. Stoll. Berlin, Springer.

[3] Verveen, A. A. (1960). 'On the fluctuation of the threshold of the nerve fibre. *Structure and Function of the Cerebral Cortex.* D. B. Tower and J. P. Schadé, Editors. Amsterdam, Elsevier (p. 282).

[4] McCulloch, W. S. (1960). 'The reliability of biological systems'. *Selforganizing Systems.* M. C. Yovits and S. Cameron, Editors. London, Pergamon Press.

[5] Weiss, P. (1959). *Quotation.* R. Gerard, Editor. Symposium on Concepts of Biology. New York, National Academy of Sciences.

[6] Weiss, P. (1962). From cell to molecule. *Symposium on Molecular Control of Cellular Activity.* New York, Wiley.

KEY PAPERS

[7] Matthysse, S. W. (1959). *Thesis*, Princeton, Princeton University Press.

[8] Ashby, W. R. (1950). The stability of a randomly assembled nerve-network. *Electroenceph. clin. Neurophysiol. 2*, 471-482.

[9] Cragg, B. G. and Temperley, H. N. V. (1954). 'The organisation of neurons: a cooperative analogy'. *Electroenceph. clin. Neurophysiol.*, *6*, 85-92.

[10] Beurle, R. L. (1956). Properties of a mass of cells capable of regenerating pulses. *Phil. Trans. R. Soc. B, 240*, 55-95.

[11] Lashley, K. S. (1929). *Brain Mechanism and Intelligence, a Quantitative Study of Injuries to the Brain.* Chicago, University of Chicago Press.

[12] Block, H. D. (1962). The perceptron: a model for brain function, I. *Rev. Mod. Phys., 34*, 123-135.

[13] Uttley, A. M. (1961). *Progress in Biophysics II.* London, Pergamon Press.

[14] McCulloch, W. S. and Pitts, W. (1943). A logical calculus of the ideas immanent in nervous activity. *Bull. math. Biophys., 5*, 115-133.

[15] Wiener, N. (1948). *Cybernetics.* New York, Wiley.

[16] Von Neumann, J. (1951). 'The general and logical theory of automata'. *Cerebral Mechanism in Behaviour.* L. A. Jeffress: Editor. The Hixon Symposium. New York, Wiley (p. 1).

[17] Von Neumann, J. (1956). 'Probabilistic logics and the synthesis of reliable organisms from unreliable components'. *Automata Studies.* C. E. Shannon and J. McCarthy, Editors. Princeton, Princeton University Press (p. 43).

[18] Allanson, J. T. (1956). The reliability of neurons. *Proc. 1st Int. Congr. Cybern., Namur,* 687-694.

[19] Muroga, S. (1960). *Rome Air Development Center.* Technical Note 60-146.

[20] Verbeek, L. A. M. (1962). 'On error minimizing neuronal networks'. *Symposium on Principles of Self Organization.* London, Pergamon Press (pp. 121-133).

[21] Uttley, A. M. (1954). The classification of signals in the nervous system. *Electroenceph. clin. Neurophysiol., 6*, 479-494.

[22] Roy, A. E. (1960). On the storage of information in the brain. *Bull. math. Biophys., 22*, 139-168.

[23] Cowan, J. D. and Winograd, S. (1963). *Reliable Computation in the presence of noise.* Boston, M.I.T. Press.

[24] Shannon, C. E. (1948). 'Mathematical theory of communication'. *Bell Syst. tech. J., 27*, 379-423; 623-658.

DISCUSSION

Wiener: There are several things here that interest me very much. Firstly, the algebra-of-logic, the yes or no of the nervous system. It looks superficially as if the simple stage of nervous

162

activity were a yes or no stage. It need not be. When we have the all-or-none law in the nervous system, as it is usually given in the case a spike propagates itself or the spike disappears, which are usually regarded as the only possible alternatives, that is not a fair representation of what happens. It takes some distance along the neuron for the spike either to assume a stable form or to disappear. These are with respect to the individual neuron long-time phenomena. The importance of this is that the element, the elementary action, in the neuron is not the algebra-of-logic action given here, but a much more complicated action. In the short fibres of the cerebrum it is at least highly questionable whether the runs are long enough to give such a definite yes or no. In other words, when we go to the elementary machine in the nervous system, the degree of complexity of the nervous system is not given simply by the number of neurons, but by the number of stages of activity in each neuron. This is enormously greater, and this may be very important.

The other thing I want to speak about is the statistical mechanical implication. I am quite convinced too that this work is very closely allied to general statistical mechanical work. For that you will have to consider non-linear changes in the neuron. These changes need not be of the all-or-none nature. This all-or-noneness in that case as well is a result of a simplification. This sort of work will be most necessary in long distance forces, with plasmas for example. I am working now on a form of statistical mechanics for that. What is relevant in this connection is that the model I am using does not go back to discrete particles, but goes back to a continuum in which the number of degrees of freedom effectively is kept finite, not by the existence of individual particles, but by certain limitations on the harmonic analysis in space. I feel that this sort of work is going to be useful also in the discussion of nerve nets. That is, I think we need a thorough theory of random functions here and I think that while pure atomism both in neurology and in particle physics is itself one way of approach which has a great deal of validity and which we cannot afford to supersede completely, there are other formalizations which may be equally useful or more useful in many cases.

Ten Hoopen: The mathematics of these nets of formal neurons is very interesting indeed. However, I am bothered by the last sentence of your abstract: 'The intention to consider the application of these results to the construction of mathematical models of cortical structure'. Do you mean with this what you said about Paul Weiss' conjectures? This is asked in view of

163

what Dr. McCulloch stressed: you are using idealized, formal neurons. Could you say something about the relation between these neurons and those actually found in the cortex?

Cowan: In our own group Dr. Lettvin has argued that it is illogical to attempt to apply automata theory to nerve nets. He argues that there exist units which have as many as 100,000 inputs and 1 output ramifying to many places. The geometry of the units is all important and there may be up to 10,000 different kinds of units in the nervous system. Furthermore, he argues that all kinds of slow potential changes occur within such a unit, so that one cannot in fact talk of it as an automaton. But yet when it comes down to a final analysis, that unit with its 100,000 discrete inputs and 1 output, is a finite automaton. It does not matter what is going on inside, i.e. what kind of slow potential changes are occurring, you have got discrete impulses as inputs. It is true they get transduced into all kinds of slow potentials and thus all kinds of electrochemical effects occur, however, the final output again is a propagated action potential, and patterns of impulses emerge from the unit. It therefore seems to me that one can still apply statistical theories of automata to nerve nets. Statistical methods are of course extremely weak for such an application, and will not work for specific cases. You have to put this specificity in. But such theories certainly might tell one something more about the qualitative behaviour of a large mass of interacting units, that exhibit integrated behaviour.

George: Do the neural net systems you have, make in any way an attempt to describe *growing* processes? Do you have in mind an *adult* neural net?

Cowan: We are not even as specific as that. Some work has been done at the Carnegie Institute of Technology by W. H. Pierce on adaptive systems which can learn to correct their errors by finding out which inputs are less reliable.

George: These prewired systems could not simulate growth as they stand.

Cowan: As they stand, no, but it would be very easy to introduce growth if one had sufficiently complex units.

George: It occurs to me, not too seriously perhaps, that the girls who wire the network to whom you ascribe an error probability function, represent the growth process of your machine. If these girls were replaced by machines, the constants on the machine may be of real interest.

Bok: I would like to ask a question. In your system you have put in something, you get out something and now your idea is that you have certain laws between the input and the output.

Why don't you take the input only and use that input to ask the question: what is the function of our nervous system?

Cowan: I should have probably stated at the start that this analysis is really secondary because one is assuming that one already knows what one wants to do, that one has a blueprint for an ideal machine. Much more important questions concern the possibility of actually duplicating the given natural components that we have and of actually discovering what the random system is doing, but that is not what we have attempted here.

Rashevsky: I would like to make a few remarks. Can it be that nature acts in some way as an engineer would probably act, without attempting to explain how and why the whole thing happens. After all, in physics we accept postulates and principles which we cannot visualize and explain. Light is both a wave and a particle and this caused many headaches to the physicist some 35 years ago but now we just accept this idea and do not analyse it further. It may be that we will have to accept that the brain was designed by a very smart engineer, and we might not be able to find why it happens so. On the other hand, there may be an evolutionary principle involved allowing the better designed system to survive. We cannot actually conclude that this is an evolutionary process, until a thorough mathematical investigation of certain other aspects of the evolution is made.

Hoogenstraaten: Does your proposed method, for eliminating the non-reliability of components and/or wiring and for obtaining in this way a reliable assembly, remove the contradiction met with when considering finite and *strictly causal* machines (with infinite tape), giving correct answers to logical propositions. With such machines, there can always be found a proposition which is intuitively true, but which cannot be proved by the machine. Identification of our brain with such a computer would give rise to a contradiction.

Cowan: I am not sure if I understand your question exactly, but there are certain features of complex machines like this that are associated with an indeterminacy. Suppose you had a machine that is actually constructed so that it controls malfunctions and 'survives' in a noisy environment. A necessary property of such a machine is that one unit interacts tremendously with many others, and any one unit performs some composite of many of the things that are necessary for the survival of the system. The logical description of the input-output relations of this system requires only two-valued Boolean logic. But the logical description of relations within the system requires many-valued logic. One only goes from the many-valued to the two-valued logic when one goes from the inside of the system to the outside. And in fact given knowledge of input-output relations outside, one cannot infer anything about the inside, and the output of a single neuron (i.e. something inside) does not tell you anything about the system either. So there is a basic uncertainty in the whole business between what is inside and what is outside.

Stewart: This sort of theory seems to assume that at any time there will be more or less constant fractions of faulty components. There seems to be then an implication that a component may at some time fail but it really does not stay failed. Did you look into cumulative failures?

Cowan: It turns out that if you go to sufficiently large machines it does not matter too much whether the components malfunction intermittently or whether they fail permanently, provided only that such errors are not catastrophic.

Stewart: I mean, let us assume that something fails and once it has failed stays that way, is that perhaps then a little better model of the way some things occur in living organisms.

Cowan: If all failures are permanent and go on increasing, you cannot do very much about it, except using more and more components or in fact using replacement techniques. You need a kind of dynamic sort of process to take care of that.

Coldacre: I am interested in what you said about the ability of your system to tolerate errors in its internal connections, and I was wondering whether you could give a numerical estimate of this in particular instances. For example, how big would your system need to be to tolerate, say, 50 per cent of errors in its internal connections, and is there any possibility of a large system working with entirely random connections?

Cowan: Yes, but the computer would have to be indefinitely large, and it would depend very much on the actual noise of the components.

Uttley: It is not just the size of the machine which comes into question, but how complicated the computing elements are.

Cowan: Let me give the following example. Suppose errors of interconnection occur with a probability of one-half, and suppose components are available, which have a computation capacity of $0·9$ (i.e. $0·1$ bits of output information are lost due to noise). Then we have the result that the redundancy R, that must be used in an automaton if errors of connection are to be safely ignored, is given by the inequality

$$R > \left(C_c - p \log \frac{1}{p}\right)^{-1}$$

where $C_c = 0.9$ and $p = 0.5$ (see Cowan and Winograd, 1963). This results in the value $R = 2.5$. That is, provided every $5n$ components do the job of $2n$, then for sufficiently large n, an automaton may be constructed that functions reliably, despite errors of connection occurring with probability 0.5. Thus a large enough system could work with entirely random connections. However, the components of such a system would have to be very complex, computing functions of very many inputs, with a computation capacity of 0.9, otherwise the above result would not follow.

11

A THEORETICAL TREATMENT OF FUORTES'S OBSERVATIONS UPON ECCENTRIC CELL ACTIVITY IN LIMULUS

There are two ways to approach understanding of a complex problem: the first is primarily experimental and consists of assembling as much data as possible about the system in question and then attempting to form an hypothesis consistent with the known data. The second is to make a model of the system, either in mathematical or in physical terms, and to look for similarities between the behaviour of the model and that of the system. Both are valid and indeed can merge into one another.

In this paper W. A. H. Rushton effectively combines the two approaches, showing that data from Fuortes's experiments on the relation between stimulus strength and output in the eccentric cell of the Limulus ommatidium follow naturally from the membrane properties postulated by Fuortes. Fuortes was concerned to discover the mechanism responsible for the generator potential developed across the eccentric cell membrane during light stimulation of the Limulus eye. He showed that membrane resistance decreases during light stimulation, but not during the passage of current from a local electrode, which never-the-less, like light, evoked the discharge of action potentials. He therefore supposed that the generator potential arose as a result of a change in membrane permeability due to the action of a transmitter released from the photo-receptors during illumination. His experiments led him to suggest that the output of the eccentric cell, expressed as impulse frequency, was a unique function of the membrane potential whatever its source. Rushton shows that this conclusion is justified and to be expected.

Further Reading

Fuortes, M. G. F. (1959). 'Initiation of impulses in visual cells of Limulus'. *J. Physiol.* **148**, 14-28

Kernell, D. (1965). 'The adaptation and the relation between discharge frequency and current strength of cat lumbosacral motoneurones stimulated by long-lasting injected currents'; *Acta physiol. scand.* **63**, 409

Reproduced from *J. Physiol.* 148 (1959) 29-38 by courtesy of the Editorial Board.

A THEORETICAL TREATMENT OF FUORTES'S OBSERVATIONS UPON ECCENTRIC CELL ACTIVITY IN LIMULUS

W. A. H. Rushton

In the foregoing paper Fuortes[1] has presented an extensive experimental analysis of the activity of the eccentric cell in the eye of *Limulus* when excited by the light and by depolarizing currents. His work involves six quantities: the light I and the current i which can be varied at will, the recorded potential U and the impulse frequency n which can be measured, and the membrane resistance R and potential V which can be deduced.

The eccentric cell was impaled by a micropipette used for recording potential changes and also for passing currents. In the absence of applied current the record shows the membrane potential V *(Figure 1)*, but when current flows the large drop in potential through the (50-100 MΩ) resistance of the electrode will appear superimposed upon the record. In the experiment this instrumental artifact was balanced out by the bridge arrangement shown in *Figure 1*, in which the setting was performed as follows. First the compensation voltage was adjusted to a positive value V_0 which balanced out the negative membrane potential V when in the dark and with no current i flowing, so that $V + V_0 = 0$. Then the bridge resistance R_1 was adjusted (with the preparation still in the dark) so that there was no recorded potential U whatever the size or direction of the current pulse i (generated by S) within the range employed. If R_0 is the membrane resistance of the eccentric cell in the dark and $R_0 + R_e$ the total resistance in the arm 4 of the bridge; this setting of R_1 secures that the recorded potential U measures the potential developed in arm 4 of the bridge *minus i ($R_0 + R_e$)*. Thus in any condition of excitation.

$$U = V + V_0 + i \cdot R_e - i(R_0 + R_e), \qquad \dots \dots (1)$$
$$\therefore U + i \cdot R_0 = V + V_0 \; .$$

FUORTES'S POSTULATE

Since the technique employed does not permit either R_0 or V to be directly measured, some further relation must be given before equation (1) can be solved. Fuortes puts forward the following postulate. 'The impulse frequency n is uniquely dependent upon the membrane potential V'. This means that no matter by what combination of light and depolarizing current the membrane potential has reached some value V_1, n will always have the same corresponding value n_1. So equation (1) becomes the double equation

$$U + i \cdot R_0 = V + V_0 = f(n). \qquad \dots (2)$$

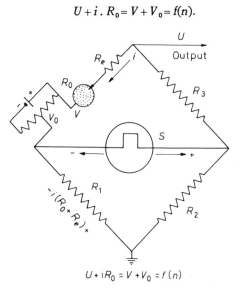

$$U + iR_0 = V + V_0 = f(n)$$

Figure 1. Fuortes's bridge balance circuit. V, membrane potential in any condition of stimulation; V_0, compensating voltage adjusted in the dark; R_0, membrane resistance in the dark; S, pulse generator to produce depolarizing current $+i$ or hyperpolarizing current $-i$.

Dr. Fuortes has kindly sent me the measurements from which his *Figure 7* was constructed and they are in part reproduced here in Table 1, which shows, for various light intensities I, and depolarizing currents i, the values of U, and n observed. If we knew the value of R_0, then from equation (2) we could test Fuortes's postulate by plotting n against $(U + i \cdot R_0)$ and finding whether the curves obtained with each value of I are identical. This is in fact done in *Figure 3*, but first the appropriate value of R_0 must be found.

In *Figure 2* the left vertical scale gives the recorded potential U and the right scale the depolarizing current i. In Table 1 the first column of the figures gives a set of values of U when $i = 0\cdot8$; and in *Figure 2* these results are represented by the pencil of lines going from $0\cdot8$ on the i-scale to the various tabulated values on the U-scale. Each line is labelled on the right with the corresponding impulse frequency n (from Table 1) and on the left with its I-value. The same is done with the second column, which generates the pencil of lines from 0, —— and so on. Now consider any ray (whose frequency label may be called n_1) and trace it as it crosses the rays of a different pencil. The first rays to be crossed will in general have n-numbers less (or greater) than n, and the last rays to be crossed will have n-numbers greater (or less) than n_1. So there is some region in the pencil where n_1 would meet another ray labelled n_1. In *Figure 2* a dot is placed upon each ray where it is judged it would meet rays of the same label. If Fuortes's postulate were true, the experiments accurate, and the points well judged, they would all fall upon a vertical line whose position determines R_0.

It is seen that in fact they lie very close to the vertical line, drawn, and the corresponding resistance R_0 is found from the dotted lines which cross upon this vertical line. They show that a fall in i of 1 nA, together with a rise of 5·5 mV, leaves $f(n)$ unchanged; so we see from equation (2) that $R_0 = 5\cdot5$ MΩ.

It will be unnecessary to give the proof of the projective properties stated above for the object is simply to arrive at the

TABLE 1. The measurements from which Fuortes's *Figure 7* was plotted

I	i	0·8	0	−0·8	−2·3	−4·3	−6·4
100	n	60	49·5	31·2	5·4	—	—
	U	10·2	11·6	13·6	16·6	19	22·9
	$U + 5\cdot5i$	14·6	11·6	9·2	+4·0	−4·7	−12·3
25	n	50·7	37	20	—	—	—
	U	8·0	9·0	10·7	13·1	16·2	19
	$U + 5\cdot5i$	12·4	9·0	6·3	+0·5	−7·5	−16·2
6·2	n	41·6	23·2	9·2	—	—	—
	U	6	6·7	8·4	10·2	12·9	15·5
	$U + 5\cdot5i$	10·4	6·7	4·0	−2·4	−10·8	−19·7
1·5	n	33	14	3	—	—	—
	U	4·3	5·2	6·5	7·6	10	11·6
	$U + 5\cdot5i$	+8·7	5·2	2·1	−5·0	−13·7	−23·6
0·4	n	28	8·6	—	—	—	—
	U	3·1	3·6	4·9	5·1	6·8	8·4
	$U + 5\cdot5i$	7·5	3·6	0·5	−7·5	−17	−26·8
0·1	U	1·5	2·0	2·8	3·2	4·6	5·4
	$U + 5\cdot5i$	5·9	2·0	−1·6	−9·4	−19·1	−29·8

value of R_0 (if any) which will permit Fourtes's postulate to be substained; and *Figure 3* will provide *a postiori* justification for the value 5·5 MΩ found. In Table 1 the numerical values of $U + 5·5 \times i$ are given, and in *Figure 3* these are plotted (horizontally) against the corresponding frequencies n, the different symbols referring to the different intensities of light. It is seen that all the points lie approximately upon the same curve, which happens to be a straight line. Thus Fourtes's postulate is substantiated and R_0, the membrane resistance in the dark, is 5·5 MΩ.

Turning back to equation (1) we know now both V_0, the compensating voltage (which was 45 mV in the experiment of Table 1),

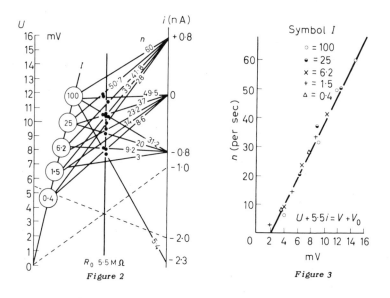

Figure 2. Vertical scale of U on left, of i on right. Corresponding values of U and i in Table 1 are plotted as lines joining these values on the scales. The corresponding values of impulse frequencies n (Table 1) are indicated at right of each line and the light intensity I at the left. If the analytic structure in this paper is true, all lines of the same n value converge upon the middle vertical line which is a uniform scale in n. And all lines of the same I value converge upon the line joining $U = 0$ and $i = 7$.

Figure 3. Relation between impulse frequency n vertically and membrane depolarization $V + V_0 = U + iR_0$ horizontally, where $R_0 = 5·5$ MΩ, from *Figure 2*. The relation for each light intensity I is plotted with a different symbol, and each relation falls upon the same curve which in fact is linear.

172

and $(U + i \times R_0)$; so we may plot the membrane potential V for any combination of light and depolarization, from the results of Table 1. This is given by the circles of *Figure 4* which is essentially a copy of Fuortes's *Figure 7*; but the ordinates are now seen to be in fact the membrane depolarization measured upwards from the resting value of −45 mV. For each intensity of light I the circles lie upon a straight line, and all the I-lines meet at a depolarization of 7 nA $(= 7 \times 10^9$ A). If R_1 is the membrane resistance in light of intensity I, then by definition

$$R_I = \frac{dV}{di} \qquad \ldots (3)$$

But from *Figure 4*, dV/di is the slope of line I which is constant throughout its length; so from equation (3), R_I is constant for a given light independent of i, but diminishes with increase in I. This marked distinction between the effect of light and of current upon the membrane resistance has already been emphasized by Fuortes.

MEMBRANE SCHEMA

Fuortes suggests in his *Figure 11* a model membrane which is exactly consistent with the above analysis and predicts that the set lines in *Figure 4* should be concurrent. The schema is reproduced here (*Figure 4A*), whence it is seen that

$$i = \frac{V - E_g}{R_g} + \frac{V - E_r}{R_r},$$

where R_r is a fixed resistance but R_g depends upon the light intensity I. V is obviously linear with i and the slope of the lines will depend upon R_g and hence upon I. For each value of V, i will in general be different lights, but when $V = E_g$

$$i = (E_g - E_r)/R_r$$

for all values of I; and thus all I lines must pass through this point. The concurrent pencil of *Figure 4* thus follows from the schema of *Figure 4A*.

173

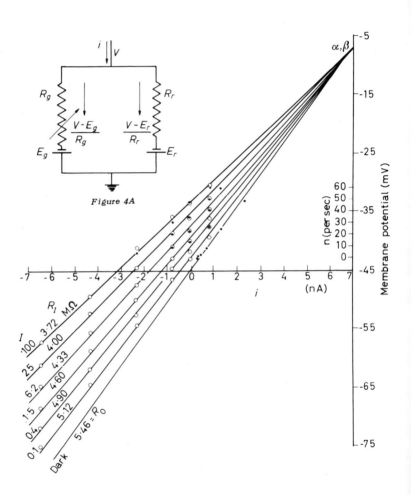

Figure 4A

Figure 4. Circles show the membrane potential V plotted against depolarizing current i. Dots show impulse frequency n plotted against i. For each light intensity I the relation is linear and all the lines meet at $i = 7$. The slope of the line gives the membrane resistance R for that light. It varies with I but not with i.

Figure 4A. Membrane schema which will account for the lines of *Figure 4* if R_g is light-sensitive, but R_r is not.

Frequency of Impulses

The relation between n, the frequency of impulses, and V the membrane potential, is plotted in *Figure 3*. So far we have only been concerned to note that the relation was a unique one, being independent of I, hence justifying Fuortes's postulate. But it is of interest to observe that the relation is linear. This is also seen very exactly in the experimental results of Fuortes's *Figures 3* and *9*.

Hodgkin[2] depolarized the unmyelinated nerves of *Carcinus* and found that many of these gave repetitive discharges of steady or very slowly-adapting frequency. Though his paper does not draw attention to the linear relation between depolarizing current and discharge frequency, this may be plotted from the records of his Plate 2, and may be seen to hold rather well up to about 60/sec. The linear relation therefore between membrane potential and impulse frequency is not a special property of the eccentric cell, and may well be a common property of unmyelinated nerve. Since there is a fixed linear relation between V and n, the scale of V in *Figure 4* is also linear scale of n. Thus the values of n plotted from Table 1 should lie upon the same concurrent pencil of lines as does V. In *Figure 4* the frequencies are plotted as dots, and are seen to fit the lines about as well as the circles do. Since discharge frequency cannot be negative, the dots cease when the I lines fall below the zero of the n scale.

This frequency plot is only a different way of representing the relation already considered in *Figure 3*, but it is the way it is represented in Fuortes's *Figure 7* and describes more explicitly his results.

In an extensive set of measurements such as those of Table 1 there is bound to be some experimental inexactness, and hence some arbitrariness as to the 'best' lines to draw. I have plotted Fuortes's points, but have drawn my own lines, which deviate slightly from his.

In particular, he obtains a value of 4·7 MΩ for R_0 by considering the relation between i and n in the dark, which we have not mentioned. These results are in fact plotted in my *Figure 4* and lie close to the line $I = 0$ (dark), but instead of coinciding with that line and giving a resistance of 5·5 MΩ they are seen to have a slope of about 4·7 MΩ, as Fuortes found. In the legend to his *Figure 7* he states that these measurements were obtained some time later than the rest of the results, so possibly by then the membrane resistance had fallen slightly.

If in *Figure 4* the concurrent point of the pencil has co-ordinates α, β, and if the line I cuts the vertical axis through $i = 0$ at V

which corresponds to n_I, the membrane resistance R_I which is the slope of the line, is given by

$$R_I = (\alpha - V_I)/\beta \qquad \ldots (4)$$

Thus R_I is a linear function of V_I, which is a linear function of n_I. This means that light changes the membrane resistance, the membrane potential and the frequency of discharge in such a way that they remain in fixed linear relations with each other.

THE EFFECT OF LIGHT UPON THE MEMBRANE RESISTANCE

The eccentric cell does not contain visual pigments, so the primary photochemical reaction must occur elsewhere, and start a train of events by which this cell is later excited. It might be thought that some electric current generated by structures near the eccentric cell flowed through it and stimulated it as did the depolarizing currents in Fuortes's experiments. But, as he emphasizes, this cannot be the case, for light stimulation is associated with a drop in the membrane resistance R_I, whereas stimulation by depolarizing currents is not. The action of the photoreceptors is therefore not to pass current but to lower the resistance R_I, presumably by releasing some transmitter substance. The relation between I and R_I, is seen in *Figure 4* by measuring the slopes of the various I lines.

In *Figure 5* these experimental values of $R_0 - R_I$ plotted against $\log I$ are shown as dots, the left-hand vertical scale giving the absolute value of $R_0 - R_I$ observed (and also the value of R_I since $R_0 = 5 \cdot 5$ MΩ). The continuous curve is the mathematical function $\log_{10}(1 + I)$ plotted (right-hand scale) against $\log I$ and slid horizontally to fit the points.

The good fit signifies that

$$R_0 - R_I = \frac{1}{2}\log_{10}(1 + 25I), \qquad \ldots (5)$$

where the number 25 of course depends upon the light units employed. It is a familiar fact that in measuring visual performance of many kinds at different levels of illumination, the performance appears to be linearly related to the logarithm of the light. This naturally does not apply to very weak lights, where the performance is around zero but where the logarithm is

$-\infty$, and an appropriate function is

$$a \log(1 + I/b), \qquad \qquad \text{.... (6)}$$

which at high illuminations is linear with $\log I$, and is zero when $I = 0$.

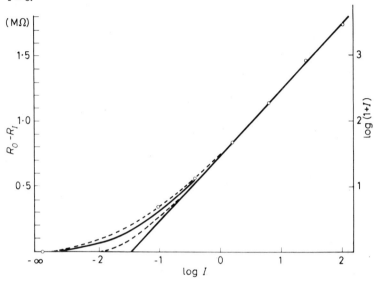

Figure 5. Points plot relation between R_I (*Figure 4*) and $\log I$. The curve is the mathematical function $\log(1 + I)$ plotted against $\log I$ and slid horizontally to fit the points. The upper interrupted curve is Stiles's increment threshold relation, $\log \Delta I$ versus $\log I$ in man: the lower is the contribution of photo-receptors to the excitation pools in the frog's retina as a function of $\log I$.

Barlow[3] has pointed out that if the retina and neural mechanism had inherent activity indistinguishable from the activity aroused by light, and of magnitude b, then this must be included with the light in any transformations, signal-to-background discrimination, etc., which are subsequently made. This leads to expression (6) if b represents the spontaneous activity occurring in the mechanism (presumably the photoceptor) where the logarithmic transformation occurs.

The upper dotted curve in *Figure 5* shows Stiles's[4] log increment threshold relation plotted against $\log I$ for the various visual mechanisms of the human eye. The lower interrupted curve represents the contributions of photoreceptors to the excitation pools in the retina of the frog (Rushton[5]). Both these latter curves are derived from records of events at least one synapse more central than the records of Fuortes's which form the basis of this paper. But since in *Limulus* the membrane resistance is related linearly to the membrane potential (equation 4), and hence to the frequency of nerve impulses (*Figure 3*), the relation of *Figure 5* may well underlie retinal interactions of quite distant and extensive kind, whether these are by the addition of '*V*'s through electronic spread or of '*n*'s by summation of impulses.

If indeed we suppose that the transmitter substance is liberated by the photoceptors in an amount proportional to

$$\log(1 + I/b),$$

and that the membrane resistance R_I falls in proportion to the amount liberated, we get exactly the relation of *Figure 5* (though it is not easy to see how such a relation can arise physically). But, further, R_I will be a linear function of the impulse frequency and each of those impulses will liberate its drop of transmitter to the synapse, so the effect will be to reproduce a synaptic change at the central end of the nerve, which is a linear function of the change at its periphery. This transmitter might be excitatory or inhibitory, and an example of inhibition which exactly fits this scheme has been demonstrated in the eye of *Limulus* itself, by the elegant experiments of Hartline and Ratliff[6].

CONCLUSIONS

In the present paper the studies of Fuortes upon exciting the eccentric cell of *Limulus* with a combination of light and depolarization have been analysed as follows:

(1) In the important relation

$$U + iR_0 = V + V_0 = kn + n_0 \qquad \dots (2A)$$

the first equation follows from the bridge circuit (*Figure 1*) used for recording, and the second is consistent with Fuortes's postulate, which states that $(V + V_0)$ is some unique function of n (the frequency of impulses) independent of the relative contributions of U and i to V.

(2) The experimental results of *Figure 3* support the postulate and show the n function to be linear. A similar function between n and i was found by Hodgkin[2] in some unmyelinated fibres in

Carcinus and it may be a common property of unmyelinated nerves.

(3) Knowing R_0 from *Figures 2* and *3*, we may plot the membrane potential V against the depolarizing current i for various light intensities I. As is seen in *Figure 4*, the results fall upon a set of concurrent straight lines whose slope is the membrane resistance R_I in those conditions. It follows that the membrane resistance is affected by light but not by current.

(4) The way in which R_I is affected by light is shown in *Figure 5*, and the good fit with the mathematical curve substantiates the relation

$$R_0 - R_I = a \log(1 + I/b). \qquad \ldots \ldots \quad (7)$$

(5) If the schema of *Figure 4A* be taken as model of the cell membrane where R_g depends upon light but R_r does not then the relation between V, i and I to be expected is a concurrent set of straight lines as those actually found (*Figure 4*). Thus the quantitative structure of the observations probably follows from this electrical structure of the cell membrane.

(6) It must, however, be noted that since R_g is the only component of the cell membrane which is light-sensitive, the relation between light and resistance (equation 7) should be expressed primarily in terms of R_g, not of R_g in parallel with R_r which is R_I. This makes the equation look much less attractive.

(7) However, R is a useful entity to employ since over the range of these investigations not only does it exhibit the simple empirical relation of equation (7), but also it is a linear function both of the membrane potential V and the discharge frequency n.

SUMMARY

1. The experiments of Fuortes upon the eccentric cell of *Limulus* presented in the foregoing paper are here analysed.

2. Fuortes's conclusions are strengthened, and his results shown to follow from the membrane structure he proposes.

3. The main steps in the analysis are summarized under ' Conclusions '.

I am very greatly indebted to Professor A. L. Hodgkin for his help in presenting the ideas of this paper.

REFERENCES

[1] Fuortes, M. G. F. (1959). Initiation of impulses in visual cells of *Limulus*. *J. Physiol. 148*, 14-28.

[2] Hodgkin, A. L. (1948). 'The local electric changes associated with repetitive action in a non-medullated axon. *J. Physiol. 107*, 165-181.

[3] Barlow, H. B. (1957). 'Increment thresholds at low intensities considered as signal/noise discriminations'. *J. Physiol. 138*. 469-488.

[4] Stiles, W. S. (1953). 'Further studies of visual mechanisms by the two-colour threshold method'. *Union Internationale de Physique pure et appliquée. Coloquio sobre Problemas Opticos de la Vision. Madrid.*

[5] Rushton, W. A. H. (1959). 'Excitation pools in the frogs retina'. *J. Physiol.* (in the Press).

[6] Hartline, H. K. and Ratliff, F. (1957). 'Inhibitory interaction of receptor units in the eye of *Limulus*. *J. gen. Physiol. 40*, 357-376.

12

SENSORY MECHANISMS, THE REDUCTION OF REDUNDANCY, AND INTELLIGENCE

The concept of redundancy is one of great importance in Information Theory and Cybernetics. The massive redundancy in human speech, for example, allows one to correctly interpret communication despite heavy interference from noisy channels. But what happens when a message, complete with redundant information, is received almost in full in the absence of interference *en route?* How much of the message is then stored? And if only partly, stored, how is the selection of the 'significant' information made? Such questions bring one straight into the vital topic of filtering and coding—a field where engineers and physiologists find themselves in particularly good communication. In this paper Horace Barlow suggests that the storage of redundant information would present problems yet envisaged and proposes some simple coding and classificatory systems which are physiologically plausible. In reading this paper it will do no harm to remember that the systematic and selective rejection of a good deal of the information gathered at receptor level is a convenience to the brain—not solely from a point of view of storage space, but also to allow the maximum speed of response.

Further Reading
Jacobsen, H. (1950). 'The informational capacity of the human ear'. *Science, N.Y.* **112**, 143
Jacobsen, H. (1951). 'The informational capacity of the human eye' *Science, N.Y.* **113**, 292

SENSORY MECHANISMS, THE REDUCTION OF REDUNDANCY, AND INTELLIGENCE

H. B. Barlow

SUMMARY

Psycho-physical and physiological investigations have shown that the eye and the ear are remarkably efficient instruments: consequently the amount of information being fed into the central nervous system must be enormous. After a delay, which may vary from about 100 ms to about 100 years, this information plays a part in determining the actions of an individual: therefore some of the incoming information is stored for long periods.

The argument is put forward that the storage and utilization of this enormous sensory inflow would be made easier if the redundancy of the incoming messages was reduced. Some physiological mechanisms which would start to do this are already known, but these appear to have arisen by evolutionary adaptation of the organism to types of redundancy which are always present in the environment of the species. Much of the sensory input is not shared by all individuals of a species (e.g. stimuli provided by parents, language, and geographical locality) so a device for 'learning' to reduce redundancy is required. Psychological experiments give indications of such mechanisms operating at low levels in sensory pathways, and 'intelligence' may involve the capacity to do the same at high levels.

In order to exemplify the operations contemplated, a device which reduces the correlated activity of a pair of binary channels is described.

The usual mechanistic approach to the higher nervous system begins with a consideration of the factors which can be shown to have an immediate effect on the output of the nervous system. The commonest starting point is the simple monosynaptic reflex in which a single sensory input controls a single motor output, as shown diagrammatically in *Figure 1(a)*. The next stage is to elaborate this by taking into account other sensory modalities,

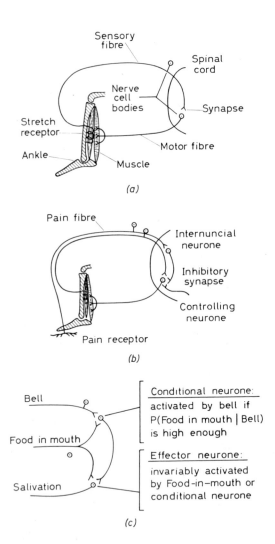

(a)

(b)

(c)

Figure 1. Diagram showing approach to higher nervous function from motor (effector) side. *(a)* monosynaptic stretch reflex; *(b)* same with addition of internuncial neurones, controlling neurones from other parts of the central nervous system, and inhibition by pain endings; *(c)* conditioned reflex.

184

inhibition, internuncial neurones, and controlling neurones from elsewhere in the nervous system, as shown in *Figure 1(b)*. With all its trimmings this gets one to a stage of complexity perhaps comparable to that of an automatic tracking radar set, or the automatic pilot of an aeroplane. It will show none of the plasticity or adaptability to new surroundings which is characteristic of the higher nervous system, so the Pavlovian conditioned reflex is next introduced. The principle here is that if there are two sensory stimuli (Bell and food in mouth), one of which (food in mouth) always produces a response (salivation), then if they occur jointly with sufficient frequency, the one which, to begin with, did not cause a response, begins to do so (Bell alone causes salivation). This is shown diagrammatically in *Figure 1(c)*, and is perhaps the simplest type of learning behaviour that has been studied in animals, though it has not been investigated in a simple isolated preparation as the diagram might suggest. Uttley (1954, *refs. 22 and 23*) has clarified the principles of operation of such mechanisms and built conditional probability devices which show the same properties of learning and inference.

Now the simple feed-back diagram in *Figure 1(a)* has a single input channel, *Figure 1(b)* and *(c)* have two inputs, and Uttley's machine has up to five inputs; but a human brain has something like 3×10^6 sensory nerve fibres leading into it. If it could be supposed that a million or so devices like that of *Figure 1(c)* would deal with the sensory inflow one would be well satisfied with the understanding gained from this approach: but this is not so. The essential operation in a conditional probability device is to measure the frequency of occurrence of *combinations* of activity in the input. Now if the number of binary inputs is increased from two to a million the number of possible combinations is increased from 2^2 to $2^{(million)}$; an arrangement like that of *Figure 1(c)* takes one less far than at first sight appears. I think it follows from this consideration that conditional probability machines cannot be fed with raw sensory information, and the problem of digesting or processing the sensory information entering the brain is an important one. Furthermore, modern electrophysiological techniques are making it possible to record from nerve cells at various levels in the sensory pathways, so this is a problem which is becoming accessible to experimental investigation.

In this paper I have first tried to make rough estimates of the rate at which information flows into the human brain. It is then suggested that an essential step in organizing this vast inflow is to derive signals of high relative entropy from the highly redundant sensory messages. For this something similar to the optimal codes discussed by Shannon[19] needs to be devised for the sensory input, and the steps required to do this are considered. Finally, a modified form of such recoding is proposed, some evidence that it occurs is brought forward, and it is suggested that the idea may be extended to cover some of the processes going on in consciousness and called reasoning or intelligence.

1. THE SENSORY INFLOW

(a) Properties of Nerve Fibres

We are equipped with sensory instruments of astonishing sensitivity and versatility which supply information about the environment to the central nervous system. This information is carried along nerve fibres, and since a good deal is known about what these fibres can and cannot do, one can derive an approximate upper limit to the rate at which information enters the brain. If the simple assumptions are made that (i) the maximum frequency of impulses is 700/s, and (ii) in 1/700th. s a nerve can only be used to indicate the presence or absence of an impulse, then the maximum rate at which it can transmit information is 700 bits/s. Mackay and McCulloch[16] point out that the nerve might be used more efficiently if, instead of detecting the presence or absence of an impulse, the intervals between impulses are used to convey information. Using such pulse interval modulation, and assuming (i) accuracy of estimation of intervals of 0·05 ms, (ii) a minimum interval of 1 ms, they give the maximum capacity as 2,880 bits/s. This would require a mean frequency of 670 impulses/s, but at a mean frequency of 50/s, such pulse interval modulation still allows 500 bits/s to be transmitted. These figures are actually too low, because Mackay and McCulloch incorrectly assumed that the optimum distribution of intervals was uniform instead of exponential: however, if the other assumptions are granted, they show clearly that a single nerve fibre could be used to transmit information at a rate well above 1,000 bits/s.

The total capacity of the sensory inflow appears to be above 3×10^9 bits/s, but it is certain that nothing like the full capacity is utilized. The mean frequency of impulses must be far below the optimum; peripheral nerves appear to use pulse frequency rather than pulse interval modulation, so that there will be high serial correlations between the values of intervals; furthermore, there are generally considerable overlaps in the pick-up areas of neighbouring fibres, which are therefore bound to show correlated activity. Finally, the figure for the performance of a nerve fibre given above might be approximately true for the large diameter fibres, but those of smaller diameter, which make up a large fraction of the total number, must have a smaller capacity. It would be pure guesswork to try to allow for these factors, but one can get indications of the utilized capacity from two other sources.

(b) Sensory Ability

Jacobsen[13,14] has made estimates of the informational capacity of the ear and the eye. For the ear he calculated 50,000 bits/s from the number of discriminable pitches (about 1,450), the number of discriminable intensities at each pitch (average about 230), and the time required to make such discriminations (1/4 s). This does not make any allowance for masking—the observed fact that the presence of one tone interferes with the perception of other tones. Jacobsen calculated that this would reduce the information capacity by a factor of about six, bringing it down to 8,000 bits/s. Now there are 30,000 nerve fibres from the ear, so each fibre must carry an average of about 0·3 bits/s.

For the eye he calculated from published data of central and peripheral acuity that there were 240,000 resolvable elements in the visual field (he seems to omit a factor of two in the integration, but this is perhaps compensated by the rather high figure for acuity which he uses). He supposes that each element can be discriminated at two intensities, with an average temporal resolution of 1/18 s. These figures give $4·3 \times 10^6$ bits/s. In the optic nerve there are just under a million fibres, so about 5 bits/s are conveyed on the average by each fibre.

There are crude estimates. For instance, no account has been taken of colour discrimination, or of the ability to localize a sound by binaural effect and judge depth by stereoscopic vision. Nevertheless, they are probably of the right order of magnitude and they are probably good enough to justify the claim that optic nerve fibres carry much more information than those of the auditory nerve. This may be significant and will be referred to later.

These figures suggest that total sensory inflow along the three million sensory fibres is rather under 10^7 bits/s.

(c) Communication Bandwidths

The capacity of the communication channels engineers need to transmit auditory and visual signals is clearly related to the capacity of the sensory pathways. Engineers, in the interests of economy, may be expected to try to use the narrowest bandwidths which will satisfactorily load up the sense organs involved, and recipients may be expected to insist that such satisfactory loading is not too far short of normal loading.

Ten kilocycles bandwidth at 40 d.b. signal noise ratio give a good quality auditory signal, and has a capacity of 133,000 bits/s. This is more than ten times Jacobsen's final figure for the capacity of the ear (8,000 bits/s), and the discrepancy is presumably due to (i) the transmission of relative phases of the frequency components, which gives information not utilized by the ear—at least in the type of discrimination taken account of by Jacobsen; (ii) the failure of the engineer to exploit the loss of efficiency of the ear which results from masking.

A satisfactory 400 line television picture requires three megacycle bandwidth at about 10 d.b. signal-noise ratio, and this corresponds to 1.2×10^7 bits/s. One is much more aware that such a television picture falls short of one's normal visual signals than one is in the case of a 10 kc 40 db auditory signal because it does not fill the visual field, and lacks detail and colour, but it is still more than double Jacobsen's estimate of the eye's capacity. In this case the most notable matching errors are the failure to exploit (i) low peripheral acuity of the eye, (ii) reduced temporal and spatial resolving power in low intensity regions of the image.

Engineers seem to require 5-10 bits/s channel capacity per nerve fibre to load up our sensory pathways, but the discrepancies between this figure and those obtained from direct estimates of sensory abilities can probably be attributed to poor matching.

(d) Time of Storage

Not only is the input to the nervous system enormous, but some at least, of the messages received are stored for very long periods. Most people would agree that sensory impressions can be recalled after a lapse of, say, 70 years, and sometimes a person can produce objective evidence of the accuracy of his recollections. In addition there are, of course, many sensory impressions which cannot be recalled, but which have, none the less, left their mark: we do not remember the successes and failures by which we acquired the correct usage of 'yes' and 'no', but this correct usage is often retained beyond the retiring age. If one allows for 50 years of waking life, the total sensory input is something like 10^{16} bits. Complete storage of all this information is neither likely to be possible nor, of course, is it what is needed.

(e) *Fate of Sensory Information*

The rest of this paper is about a suggested plan of storing and displaying this enormous sensory input, but one must first have some idea of the use that is made of the sensory information and the neural equipment which is available for dealing with it. According to Craik[4] the sensory information is used to build up a model of the external world which provides a basis for determining what course of action is most likely to lead to the survival of the individual and his species. That is a brief answer to the first question, and it also gives the answer to another fact which might otherwise be puzzling. A man can only make decisions on the basis of sensory information at a maximum rate of about 5 to 25 bits/s. (Hick[11], Quastler[18]): why, then, does he need a sensory input of 10^7 bits/s? Craik's answer would probably have been that the greater the sensory input the more complete and accurate the model, and hence the surer its basis for planning survival.

The question of the equipment available can also, because of our ignorance, be answered briefly. There are some 10^{10} interconnecting nerve cells in the central nervous system, and quite a large proportion of them must be available for the task of dealing with the sensory input and building up the model. We are only beginning to determine the properties of these cells; it has been known that their long processes transmit information as all-or-none impulses for more than 50 years, but how information is stored is not yet understood. In what follows I shall be talking about *what* the nervous system does rather than *how* it does it, so our ignorance of the method of storage of information is not too serious. The problem might be discussed abstractly, but for the sake of a definite model one can think of each nerve cell having 'excitation laws' which determine the conditions under which it becomes active, and suppose that these laws can be changed so that it becomes active in response to a different set of patterns of activity in the nerve cells in contact with it. The excitation laws for all the neurones would then form a store of information and the current display would consist of the pattern of nerve cells which are actually transmitting impulses down their long processes at any given moment.

With this model in mind the problem is: what should the excitation laws of the neurones be, and how should they be alterable, in order that the display of activity shall help the individual and species to survive in the situation giving rise to the current sensory input? To avoid basing the argument on uncertain preconceptions of what the brain does, one could put it in more general terms in this way. The barrage of nervous impulses reaching the nervous system seems to be unmanageably large; how should a selection of this activity be made for current display and future reference?

2. ORGANIZATION OF THE SENSORY INPUT

The proposition is that the initial selection is performed according to those statistical properties of the past sensory messages which determine how much information particular impulses convey. It is supposed that the sensory messages are submitted to a succession of re-coding operations which result in reduction of redundancy and increase of relative entropy of the messages which get through. Ideally one might imagine that an optimal code is constructed, so that the output, or 'display' of current input, has no redundancy, relative entropy 1, and carries all the information of the input. This ideal obviously cannot be reached, but the recoding operations are supposed to tend towards the ideal: that is, outputs are derived from the input, which have high relative entropy and carry as much of its information as possible.

Shannon has shown that it is possible in principle to obtain near optimal coding if a sufficient number of messages of a given length have occurred to give knowledge of the statistical structure of the messages, and if delays are permitted between input and output. Fano and Huffman[12] have described procedures for constructing such codes. The first steps are to define what shall constitute a single message and then to measure the frequency of occurrence of all possible messages of this class. Clearly the class cannot be the whole of the sensory input to the brain up to a particular moment, for this message has only occurred once. The input must be sub-divided in time, and first consider the operation required to re-code messages of duration, say, one second. The capacity of the input channel has been shown to be about 3×10^9 bits/s which corresponds to $10^{(\text{thousand million})}$ possible messages per second. If one takes account of the restrictions which reduce the utilized capacity to some 10^7 bits/s, and considers messages of one-tenth second duration, there are still some $10^{300,000}$ possible messages. It would clearly be hopeless to devote neural equipment to the counting of each possible message, for it is highly improbable that any single message will be exactly repeated and most of such equipment would be unused at death. This is, essentially, the same difficulty that was levelled against the idea that conditional probability devices could be served with unprocessed sensory data, but when one considers optimal coding there is a possible solution. Because the code is reversible, no information is lost by re-coding small sections of the sensory input independently, and such preliminary re-coding will enable the whole message to be passed down a channel of smaller capacity, and thus facilitate subsequent steps.

The idea is best illustrated by considering the order in which different types of redundancy might be encountered, and eliminated, during the successive re-coding operations. First there is the very large amount which results from the inefficient utilization of peripheral nerve fibres. Looking only at the nerve impulses as they arrive, it would be found that impulses occurred at different mean rates in different fibres and in all of them at rates well below the optimal frequency for information transmission. This type of inefficient utilization of a set of communication channels is a form of redundancy, but for reasons discussed later (Section 4) it may be less important to eliminate than other forms: for the moment one can consider the capacity of a nerve fibre as determined, not by maximum frequency of impulses, but by the mean frequency at which they occur.

Next, still looking only at the impulses as they reach the central nervous system, it would be found that impulses do not occur completely at random in time but tend to follow one another in sequences and bursts: the first re-coding operation might be a mechanism which reduced the serial correlations so that the same amount of information was carried by fewer impulses. In addition it would be found that certain groups of nerve fibres tended to become active at the same time. These would be fibres whose receptive fields on the sensory surface overlapped, so that this particular form of redundancy results from the anatomical properties of fibres and sense organs, just as the serial correlations in time result from the fact the intensity of a stimulus is coded as frequency of impulses at the sense organs.

These first steps, then would reduce the orderliness in the sensory messages which results from characteristics of the sensory apparatus. But if this orderliness can be eliminated, so can that resulting from the characteristics of the environment which is providing these stimuli. For instance, it will often happen that a stimulus covers more than a single point on the sensory surface and therefore causes activity in a group of fibres larger than those whose receptive fields overlap. Advantage could be taken of this to reduce the number of impulses required to convey information about such a stimulus. Again, a stimulus will often be moved across a sensory surface causing excitation in sequences of nerve fibres. Such repeated, ordered, sequences of activity would be a form of redundancy which could be reduced by suitable re-coding. In fact, any pattern of stimuli which represents a departure from complete randomness—such as simultaneous stimuli at different points on the sensory surface, stimuli which are maintained for long duration of time, ordered sequences or cycles of stimuli—present an opportunity of reducing the magnitude of the sensory inflow by suitable re-coding. It is clear that many of the complex features of our environment will come into this category. For instance, the stimuli which result from an animal's parents or its habitat are repeated frequently, and economies could be effected by reducing the space in the sensory representation occupied by these familiar stimuli and allowing more space for the infrequent and unexpected stimuli.

It is suggested, then, that the processing or organization of sensory messages is carried out by devising a succession of optimal or near-optimal codes adapted to the messages which have been received. In the early stages the total inflow will be sub-divided into many small sections, presumably taking in each section the messages coming along neighbouring fibres during a short interval of time. In the later stages the coded outputs will be re-mixed, possibly with the addition of delayed inputs (as utilized by Uttley in conditional probability devices) to allow detection of movement and other ordered sequences of activity, and then will be sub-divided again into small sections. Thus in the later stages the nerve messages being re-coded may be derived from more and more remote parts of the sensory inflow and may also come from sensory stimuli more and more separated from each other in time of occurrence. It will be seen that at each stage storage of some of the sensory information is required in order to construct the optimal code, and thus the code itself forms a kind of memory.

Now the idea that our brains detect order in the environment is not new. Empiricist philosophers have talked of percepts being associated sense impressions, and of causality corresponding to invariant succession of sense impressions. Behaviourists have emphasized the importance of association, and Gestalt psychologists talk of ordering sensation according to certain schemata (though here there seems to be some confusion as to whether the ordered schemata are derived from sensations or imposed upon them). Thus the fact that our higher centres are much concerned with the redundancy of the sensory messages has often been pointed out, but two aspects of this fact have not, I think, been so widely recognized. First, the detection of redundancy enables the sensory messages to be represented or displayed in a more compact form; and second, the reduction of redundancy is a task which can be subdivided and performed in stages. *Figure 2*

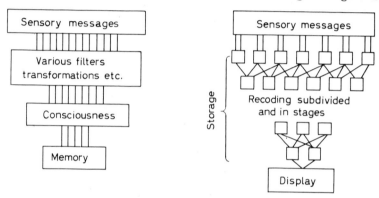

Figure 2. Diagram contrasting memory *after* consciousness in orthodox scheme with storage *before* display in optimal coding scheme

shows diagrammatically how the suggested scheme of storage and display compares with more orthodox representations of memory and consciousness. It will be seen that in the present scheme a large part of the storage of information occurs before the display—that is before the level of re-coding which might correspond to conscious awareness of sensory stimuli. The re-coding is supposed to continue at conscious levels, so some of the information reaching consciousness is also stored, but this would only be sufficient, first, to enable the process of building up the code to continue, and second to enable 'useful' association to be made between motor acts and features of the current sensory input (e.g. between salivation and bell).

It seems a help to consider the processing of sensory information as optimal or near-optimal coding for two reasons. Practically, it enables the subject to be approached along the firm path of sensory physiology instead of through the shifting sands of conscious introspection and philosophy. And conceptually it shows a way in which complete mental acts, which seem appalling in their complication and perfection, may be sub-divided into a succession of much simpler operations; this is clearly a prerequiste for gaining an understanding of the physiological basis of mental function.

It is worth noting that the possibility of sub-division rests on Shannon's proof of the possibility of near-optimal coding; if the early transformations of the sensory information were not reversible, redundant features which are detected later might be lost: and if the earlier transformations did not increase the relative entropy of the messages, they would not facilitate the detection of higher order redundancy.

3. DESIRABILITY OF OPTIMAL CODING

In the last section an outline scheme for dealing with the enormous sensory inflow was suggested. In this section some reasons for the desirability of optimal coding are put forward. It will be argued that it is desirable on the grounds of access-ibility, stability, and economy, and because it requires storage of information sufficient to form a model of the animal's environ-ment. Of course, such arguments for its desirability are not sufficient reasons for believing that it actually occurs.

(a) Accessibility

Optimal coding will improve the accessibility of information in two ways. First, the capacity of the display required for the current sensory input will be decreased. This simplifies the task of finding useful associations just as reducing the size of a haystack simplifies the task of finding needles. The second way is less obvious. In messages of high relative entropy, the probability of a given message occurring is close to the product of the probabilities of the individual signs which make it up. Now a dog feeds once or twice a day, and when looking for sensory correlates of salivation it would not be worth while to search among combinations of individual signs whose probability of joint occurrence was so low that they would be expected only, say, once a week, nor amongst those whose probability was so high that they would be expected, say, once an hour. If the input to a conditional probability device is known to be of a high relative entropy, great economies of design are possible.

(b) Stability

It is sometimes argued that redundancy is a good thing because it protects a message from noise. There may well be random effects inside the nervous system against which the storage and display of sensory information needs protection, but the redundancy of the internal representation which would achieve this is not in general the same as the redundancy which occurs in the sensory input. When driving at night the internal representation of a pedestrian crossing the road requires as much protection as the representation of the blinding glare from an oncoming car, but in the incoming sensory message the former may be represented by a barely significant disturbance in the pattern of nerve impulses, the latter by high frequency volleys of impulses in many fibres. Stability of storage and display require, at least, a re-adjustment of the redundancy of the sensory messages.

(c) Economy in Transmission and Storage

Sensory information has to be transmitted from place to place in the central nervous system and the reduction of redundancy before this is done would enable the number of internal connecting fibres to be reduced. An example where the economy so effected seems to be particularly desirable is the connection between the eye and the brain. It would clearly interfere with the mobility of the eye if the optic nerve was very much larger than it is, and according to Jacobsen's estimates it would have to be fifteen times larger if the nerve fibres were utilized as inefficiently as they are in the ear. The attainment of this 15-fold economy may, as Jacobsen suggests, be the main function of the nervous layer of the retina which links receptors to optic nerve fibres. Squids and octopuses form an interesting comparison, for they have eyes which are comparable optically to those of vertebrates, but their retina is much simpler with no synaptic layer—the optic nerve comes direct from the receptor cells. It is bulky, containing a vast number of fibres, and seems likely to be a factor restricting the mobility of their eyes.

The same argument might be applied to storage of information, since it is clearly more economical to store messages after their redundancy has been reduced. Here, however, there is a complication. The devising of a redundancy-reducing code requires storage of certain properties of the sensory message, and it has not been shown that more capacity would be saved by storing messages after re-coding than would be utilized in devising the code. The condition that this should be so depends upon the number of the times that the code, once devised, is subsequently utilized, but a discussion of this point cannot go far without knowing what parts of the sensory inflow are in fact stored: the argument of the next section is that the coder itself stores sufficient information to form a working model of the animal's environment, and therefore represents a large fraction of the total storage the animal needs.

(d) Modelling the Environment

Craik suggested that sensory information was used to form a model of the animal's environment. By a model one does not mean a simple copy of those aspects of it which have given rise to sensory stimuli: it must also mimic the structure of the environment, so that an operation performed on the model will give the same result as the analogous operation performed on the environment. When the schoolboy turns his model engine round, he receives visual stimuli similar to those he would have received if a real engine had been turned round in front of him. The model imposes restrictions on the sensory stimuli which are received in certain situations, these restrictions being the same as those inherent in the properties of the object modelled. Now it is precisely these restrictions—the departures from complete randomness of the sensory input—which the coder utilizes to increase the relative entropy of the signals. The particular code adopted is related to the particular restrictions of past sensory inputs and is therefore, in a sense, a model of the animal's environment. In the example above, the model was static, but the restrictions must often be dynamic; sets of sensory stimuli frequently follow one another in a repeated sequence, and such repeated sequences will also be reflected in the particular code adopted. Thus the code contains a working model of the environment.

If the code stores sufficient information to form a model of the environment, its potential use in aiding survival is not confined to the provision of a more compact display of the sensory input. But to make full, predictive, use of these potentialities some additional facility for getting at this stored information seems to be needed. To return to the earlier example, what facility do we have for turning round the model engine in our brain so that we can look at the other aspect?

196

4. MODIFIED RE-CODING

So far the type of optimal coding envisaged has been that described by Shannon, Fano, and Huffman, in which the output is the
smallest number of binary signals capable of carrying the information of the input. At first sight this seems to be what is needed
in the nervous system, for nerve fibres transmit all or nothing
impulses and thus seem to use a binary system. However, it has
already been pointed out that the mean frequency of impulses is
well below the optimal for information transmission even in
peripheral nerve fibres, and there is some evidence which suggests that the mean frequency is even lower in the more central
neurones (Galambos[6]). Furthermore, if the Shannon type of re-
coding was occurring, one would expect to find the sensory
pathways becoming more and more compact as the sensory
information was coded on to fewer and fewer elements. This does
occur in the retina, where some 10^8 sensory elements are connected to 10^6 nerve fibres, but as one follows the optic nerve
into the brain there is no evidence of further compression on to
a smaller number of nerves, but rather the reverse. The striate
region of the cerebral cortex which is mainly, perhaps exclusively, concerned with vision, contains some 10^8 nerve cells, in
other regions of the cortex there are about $6 \cdot 5 \times 10^9$ cells (Sholl[20])
many of which must be partially concerned with visual infor-
mation. Galambos[6] gives striking figures showing how the number
of nerve fibres available for auditory information increases as
one follows the sensory pathway from ear to cortex.

These facts do not fit in with the idea that coding in the
higher nervous system compresses information into a smaller
number of nerve fibres, and suggest that, if optimal coding
occurs, the output is not in the form of binary signals at the
optimum frequency for information transmission.

For an engineer designing a communication link, the capacity
of the channel is one of the factors under his control, and he can
effect economies by coding his signals so that they require a
smaller capacity. In the nervous system the number of nerve
fibres available for a particular task must, to a large extent, be
determined genetically. One may expect evolutionary adaptation
to have performed part of the engineer's job in selecting suitable
codes for the sensory signals, but such inherited codes obviously
cannot be adapted to the redundancy of sensory input which is
peculiar to each individual. Now although the number of nerve
cells available is probably determined genetically, the number

of impulses in the nerve cells is not, and some of the advantages of optimal coding would apply if the incoming information were coded—not onto the smallest possible number of nerve fibres each working at its optimal mean frequency—but into the smallest possible number of impulses in a relatively fixed number of nerves. This type of coding can be epitomized as *economy of impulses:* the nervous system will tend to code sensory messages so that they are represented, on the average, by the smallest number of impulses in the nerve cells available. There is an important difference between this type of re-coding and the Shannon-Fano-Huffman type; the latter does not distinguish between redundancy caused by non-optimal frequency of utilization of the individual signs of the input message, and that which is caused by correlation between signs. If impulses rather than nerve fibres are economized, mean impulse frequencies of the output will be as low as the rate of inflow of information permits, and will thus possess maximum redundancy of the first type and minimum redundancy of the second type.

A reversible coding device is described in the appendix which decreases the frequency of occurrence of a pair of binary output signs by getting rid of some of the redundancy caused by correlations between a pair of binary input signals.

5. EVIDENCE

So far, some grounds for believing that the optimal coding of sensory information would be desirable have been given, arguing from the enormous quantity of information pouring in and from rather vague ideas about what the brain does with it. In this section some of the evidence in favour of the view that it does actually occur is sketched, but this is intended to show the kind of consequences of optimal coding which may be found experimentally, and is neither a claim that it has been proved to occur, nor a critical review of the evidence for and against it. The evidence comes from a number of sources.

(a) Introspection of Sense Impressions

This is a notoriously unreliable way of obtaining scientifically valid evidence, but it is immediately accessible to everyone, so it comes first. If the hypothesis is correct, the sensory messages reaching consciousness will have been partially re-coded, and will therefore have higher relative entropy and lower redundancy than the raw sensory messages. This seems to me likely to be true of the furniture of my own consciousness, and others may feel it is true also: if, however, somebody did not agree I don't think I could persuade him by verbal arguments. More objective evidence can be obtained by looking at some messages which do *not* reach consciousness but which are known to be impressed on the sense organs. Examples of this are the shadows of the blood vessels which run on the retina in front of the sensitive elements; the fact that if distorting or inverting spectacles are worn, after some days one ceases to be aware of the distortion or inversion; adaption to the curious tone quality imposed on all sounds by the average domestic wireless set and so on. In all of these examples there are features of the sensory messages which are constantly repeated and are therefore redundant; a code which increased the relative entropy of the messages might be expected to reduce their prominence, and the fact that we cease to be conscious of them suggests that this re-coding does take place before sensory messages reach consciousness.

An experimental approach to this problem may be possible through the investigation of threshold sensations. These are perhaps the simplest elements of our consciousness, and according to the hypothesis they should tend to possess the highest possible relative entropy of a binary signal after the physical limitations of the stimulus and of the sense organs have been taken into account, and they should show a tendency to retain this property in a great diversity of stimulus conditions.

(b) From Sensory Neuro-physiology

During the past 30 years various types of relation have been observed between an applied physical stimulus and the resulting pattern of nerve impulses. Physiologists have perhaps got used to these transformations and no longer think of them as something requiring further interpretation, but possibly they can be looked upon as examples of the principle of economy of impulses: the relation between the physical stimulus and the occurrence of impulses is such that the number of impulses used to convey information about the stimulus is lower than it would be with other, more straightforward, relations.

(i) Adaptation. When a sustained physical stimulus is applied to a sense organ the nerve fibre often responds with a brief burst of impulses which rapidly decreases in frequency and is not sustained for the duration of the physical stimulus. In the left

half of *Figure 3* comparison of the trains of impulses shows the economy brought about by adaptation. But it can, of course, only be thought of as an economy when compared to a non-adapting ending, and even then only when the physical stimuli naturally applied to the sense organ are frequently of a long-sustained type. Nevertheless, where it occurs, adaptation would lead to economy of impulses and Adrian[1] suggested that its function

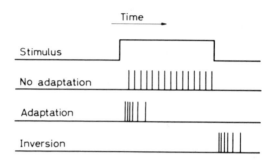

Figure 3. Diagram showing that adaptation leads to economy of impulses when a physical stimulus is of long duration, and that inversion replaces information lost by adaptation

might be to prevent an excessive number of impulses reaching the nervous system.

(ii) Inversion. It can be seen from the right half of *Figure 3* than an adapting nerve fibre fails to signal the end of a sustained stimulus. This defect could be remedied by having one which discharged as shown in the bottom line, and such nerve fibres are found. In the eye of the scallop (Pecten) Hartline[9] showed that one group of fibres discharged when a light was switched on and another group of fibres discharged at 'off'. A similar, but rather more complex, situation is found in the vertebrate eye (Hartline[8], Granit[7]). This arrangement might be thought of as making good some of the loss of information caused by adaptation.

(iii) Lateral inhibition. Adaptation increases the relative entropy of the nerve message by preventing many impulses being used to signal a physical stimulus which is constant in time. It is clear that physical stimuli will often be applied to many neighbouring receptors simultaneously, so there is a place for a spatial analogue of adaptation. The best worked out example of this occurs in the lateral eye of Limulus, where the arrangement shown in *Figure 4* has been deduced by Hartline and his

co-workers[10]. Apparently each receptor in the array exerts an influence, graded according to the number of impulses it is itself producing, which reduces the number of impulses given by neighbouring receptor units. It will be seen that the effect is to decrease the number of impulses coming from a uniformly illuminated area, while the number coming from the borders of the area are relatively unaffected. A similar situation exists in the frog

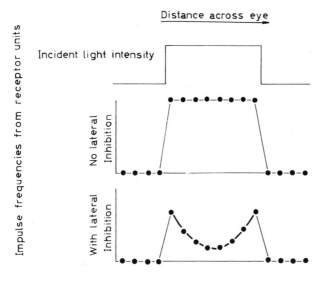

Figure 4. Diagram showing that lateral inhibition leads to economy of impulses in a uniformly illuminated area

(Barlow[2]) and cat retina (Kuffler[15]) and it has also been described in the auditory (Galambos[5]) and tactile (Mountcastle[17]; Amassian[21]) pathways.

One feature of lateral inhibition in the mammalian retina is of special interest: it is found when the retina is adapted to a uniform background light, but is absent after complete dark adaptation (Barlow, FitzHugh, and Kuffler[3]). Now it is only when the uniform background is present that the correlated discharge of neighbouring receptors will tend to occur, so it looks as though lateral inhibition is not an invariant feature of the retinal organization, but develops in the conditions where it can increase the relative entropy of the optic nerve signals. Perhaps it is a simple example of 'learnt' re-coding adapted to the redundancy which is present.

Adaptation, inversion and lateral inhibition may thus be devices used in the peripheral parts of sensory pathways to obtain signals of higher relative entropy. It is now possible to record the activity of more centrally placed neurones, but the nervous system has outwitted the physiologist who has so far been unable to determine the function of the cells whose nervous responses he has recorded. The model described in the appendix does a simple re-coding operation on two binary inputs, but it would be a difficult task to relate the output to past and present inputs without some hint about the purpose of the device. The reason, then, for putting forward the optimal re-coding hypothesis is the hope that it may be better matched to the subtlety of the nervous system than the simpler hypotheses at present entertained in physiology.

6. INTELLIGENCE

This word was added to the title in an incautious moment, but there are reasons justifying its inclusion. If it is accepted that the large size of the sensory inflow precludes its direct utilization in the control of learnt motor actions, then the mechanism which organizes this information must play an important part in the production of intelligent behaviour. In addition, when one considers the two main operations required for optimal coding there is a striking parallel with the two types of reasoning which underline intelligence.

The outputs of a code can be thought of as logical statements about the input, and, if the code is reversible, these logical statements, taken together, are sufficient to determine the exact input. Forming these statements and ensuring that they fulfil this condition are straightforward problems of deductive logic. If the code is optimal, the output statements must be chosen so that they fulfil the additional condition that, on the average, they are the smallest possible number which suffices to determine the input (for the type of modified optimal code suggested in Section 4, the additional condition is that a fixed number of possible statements are chosen for the output in such a way that the smallest number, on the average, are asserted as true). The fulfilling of these additional conditions is not exactly inductive reasoning, but it is closely related to it, for both depend on counting frequencies of occurrence of events. Having been presented with 1,000 white swans and no black ones, the relevant parts of a code would say 'henceforth regard all swans as white

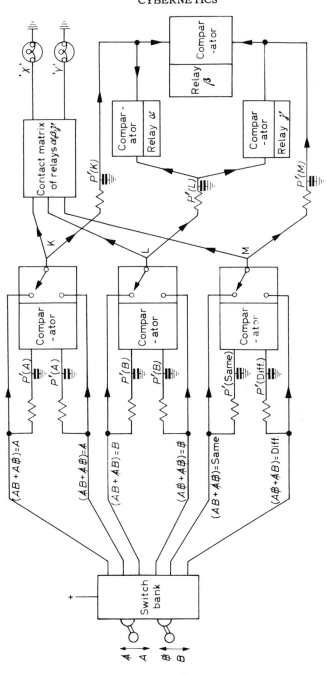

Figure 5. Circuit diagram of recoding device

203

unless told otherwise'. Inductively one would say 'all swans are white'. The tools of logical reasoning appear to be the same as those needed for optimal coding, so perhaps they can be regarded as the verbal expression of rules for handling facts which our nervous system uses constantly and automatically to reduce the barrage of sensory impulses to usable proportions.

Finally it should be made clear that the transformations of sensory messages taking place in the nervous system must, in fact, fall a long way short of true optimal coding: information must be lost, and the final 'display' must still contain redundancy. However, the fact that the image cast on the retina is not always sharp does not mean that the focusing of light by the eye is unimportant, and the suggestion is that optimal coding plays a part in the organization of sensory information comparable with image formation in the working of the eye. However, even if this conjecture is correct, the means by which it is achieved, and such matters as the classes of redundancy which are easily and naturally utilized, and the classes which are not, remain largely undetermined.

APPENDIX
(In collaboration with P. E. K. Donaldson)

Object of device. To code reversibly and without delay a pair of binary inputs *(A* and *B)* onto a pair of binary outputs *(X* and *Y)* so that the redundancy of the output due to correlations is less than the same type of redundancy in the input.

Principle used. The information carried by the inputs will, in general, be less than the capacity of the input channels *first* because of redundancy due to correlations between them *(P(AB)* $\neq P(A) . P(B))$: *second* because the frequency of signals in the individual channels is not optimum $(P(A) \neq \frac{1}{2}$ and $P(B) \neq \frac{1}{2})$. The principle used is to increase the redundancy of the second type, and so decrease that of the first type. A pair of outputs are sought which are reversibly related to the inputs, and one of which occurs with probability further from the optimum $(\frac{1}{2})$ than one, or both, of the inputs. The outputs carry the same information as the inputs, so that if such a pair can be found, the redundancy due to correlation between them must be less than is present in the inputs.

Possible codes. There are four possible input states *(AB, $A\bar{B}$, $\bar{A}B$,* and *$\bar{A}\bar{B}$),* and four possible output states *(XY, $X\bar{Y}$, $\bar{X}Y$,* and *$\bar{X}\bar{Y}$).* If the code is reversible these must be related to each other in a one-to-one manner, which can be done in 24 ways. Now since X corresponds to a pair of output states $(XY + \bar{X}\bar{Y})$, the condition for activity in X must be the occurrence of either of a pair of the possible input states, and likewise for Y. There are six such pairs: $AB + A\bar{B} = A$, $\bar{A}B + \bar{A}\bar{B} = \bar{A}$, $AB + \bar{A}B = B$, $A\bar{B} + \bar{A}\bar{B} = \bar{B}$, $AB + \bar{A}\bar{B} = (A$ and B the same), and $A\bar{B} + A\bar{B} = (A$ and B different). In addition, for reversibility, the two pairs chosen must have a common member, for if this was not so X would always be active when Y was not active, and vice versa.

After a little cogitation it will be found that there are 24 possible codes, which fall into 3 groups each containing 8 codes, the groups differing from each other in the respect which interests us, namely the division of redundancy between correlation-type and non-optimal-frequency-type. One group does not differ from the input in this respect. The other two groups do differ, and they are made up of those 16 codes for which one or other of the outputs corresponds to either $AB + \bar{A}\bar{B}$ $(A$ and B alike$)$ or $A\bar{B} + \bar{A}B$ $(A$ and B different$)$.

Condition for success, then, is that either $P(AB + \bar{A}\bar{B})$ or $P(A\bar{B} + \bar{A}B)$, should differ from ½ by more than one or other or both of $P(A)$ and $P(B)$. This is not, of course, the same as the condition that A and B are correlated, so the recoding does not always reduce correlation redundancy when this is present. Successful recoding occurs for the smallest departures from zero correlation when either $P(A)$ or $P(B)$ is close to ½.

Method. The device is made up of 6 similar units each of which compares two probabilities and operates a relay according to which is greater (see circuit diagram, *Figure 5*).

(a) Probabilities are measured by charging a leaky condenser when A (or B etc.)$= 1$; hence they are weighted for recent events, the weights decreasing exponentially with lapse of time. These time weighted probabilities are called $P'(A)$, $P'(AB + \bar{A}\bar{B})$, etc.

(b) $P'(A)$ is compared with $P'(\bar{A})$, $P'(B)$ with $P'\bar{B})$, and $P'(AB + \bar{A}\bar{B})$ with $P'(A\bar{B} + \bar{A}B)$. In each case a signal corresponding to the smaller of the pair is selected. Call these signals K, L, M.

(c) $P'(K)$ is compared with $P'(L)$, $P'(L)$ with $P'(M)$, and $P'(M)$ with $P'(K)$. Switching is performed according to the result of these comparisons so that

$$X \equiv \text{smallest of } K, L, M.$$
$$Y \equiv \text{next smallest of } K, L, M.$$

Result. The result of these operations is more specific than the original objective in that one particular code is chosen from a group of 8, any one of which would have met the requirements.

The added specificity results from the fact that we have chosen outputs which occur *least* frequently, not *most* frequently, and have arranged that $P(X)$ shall be less than $P(Y)$.

Note that if there is any logical relation in the inputs (e.g. $AB = 0$), then the outputs become mutually exclusive $(P(XY) = 0)$. If there is a double relation (e.g. $A\bar{B} = 0$ and $\bar{A}B \equiv 0$), then only one output channel operates $(P(X) = 0)$. The device might be roughly described as one which determines inductively what logical relations, if any, are obeyed by its input. If two such relations are found, one output channel is not used; if one is found, the two outputs become mutually exclusive; if none is found, but there is statistical correlation between the inputs, it will sometimes find outputs which are less correlated.

REFERENCES

[1] Adrian, E. D. (1928). *The Basis of Sensation*. Christophers, London

[2] Barlow, H. B. (1953). 'Summation and inhibition in the frog's retina.' *J. Physiol., Lond.* 119, 68

[3] Barlow, H. B., Fitzhugh, R, and Kuffler, S. W. (1957). 'Change of organisation in the receptive fields of the cat's retina during dark adaptation.' *J. Physiol., Lond.* 137, 338

[4] Craik, K. J. W. (1943). *The Nature of Explanation*. University Press, Cambridge

[5] Galambos, R. (1944). 'Inhibition of activity in single auditory nerve fibres by acoustic stimulation.' *J. Neurophysiol.*, 7, 287

[6] Galambos, R. (1954). 'Neural mechanisms of audition.' *Physiol. Rev.*, 34; 497

[7] Granit, R. (1947). *The Sensory Mechanisms of the Retina*. Oxford University Press, Oxford

[8] Hartline, H. K. (1938). 'The response of single optic nerve fibres of the vertebrate eye to illumination of the retina.' *Am. J. Physiol.*, 121, 400

[9] Hartline, H. K. (1938). The discharge of impulses in the optic nerve of Pecten in response to illumination of the eye.' *J. cell. comp. Physiol.*, 11, 465

[10] Hartline, H. K. and Ratliff, F. (1957). 'Inhibitory interaction of receptor units in the eye of limulus.' *J. gen. Physiol.*, 40, 357

[11] Hick, W. E. (1952). Why the human operator? *Trans. Soc. Instrum. Technol.*, 4, 67

[12] Huffman, D. A. (1953). A method of construction of minimum redundancy codes. Symposium on Communication Theory pp. 102-110. Editor W. Jackson. Butterworths, London

[13] Jacobsen, H. (1950). 'The informational capacity of the Human Ear.' *Science, N.Y.* 112, 143

[14] Jacobsen, H. (1951). 'The informational capacity of the Human Eye.' *Science, N.Y. 113*, 292

[15] Kuffler, S. W. (1953). Discharge patterns and functional organisation of mammalian retina.' *J. Neurophysiol., 16*, 37

[16] Mackay, D. M. and McCulloch, W. S. (1952). 'The limiting information capacity of a neuronal link.' *Bull. math. Biophys., 14*, 127

[17] Mountcastle, V. B. (1957). 'Modality and topographic properties of single neurones of the cat's somatic sensory cortex.' *J. Neurophysiol., 20*, 408

[18] Quastler, H. (1956). Studies of human channel capacity. Third London Symposium on Information Theory. pp. 361-371. Ed. by C. Cherry Butterworths, London

[19] Shannon, C. E. and Weaver, W. (1949). 'The mathematical theory of communication.' The University of Illinois Press, Urbana

[20] Sholl, D. A. (1956). 'The organization of the cerebral cortex.' Methuen, London.

[21] Towe, A. L. and Amassian, V. E. (1958). 'Patterns of activity in single cortical units following stimulation of the digits in monkeys.' *J. Neurophysiol., 21*, 292

[22] Uttley, A. M. (1954). 'The conditional probability of signals in the nervous system.' *R. R. E. Memorandum 1109*, Ministry of Supply, Gt. Malvern

[23] Uttley, A. M. (1954). 'The classification of signals in the nervous system.' *Electroenceph. clin. Neurophysiol., 6*, 479

13

DISCHARGE PATTERNS AND FUNCTIONAL
ORGANIZATION OF MAMMALIAN RETINA

There is now a considerable body of knowledge about the response properties of neurones at several levels in the mammalian
visual system. Kuffler was the first to examine the organization
of mammalian receptive fields in detail. The receptive field of a
neurone in the visual system is that area of retinal illumination
which changes the excitability of the neurone—usually measured
as an increase or decrease in the neurone's output of action
potentials. He showed that the organization of a receptive
field was quite complicated and varied considerably under
different conditions of stimulation and background illumination. Basically, receptive fields of retinal ganglion cells
contain two areas with contrasting properties. One responds to
the turning 'on' of a light and the other to its 'off'. The output of
a ganglion cell is determined by the sum of contributions from
'on' and 'off' regions of its receptive field as well as by the
activity of neighbouring ganglion cells.

The work of Hubel and Wiesel has extended our knowledge of
receptive field organization to higher levels of the visual system:
the properties of units at each level can only be understood in
terms of the properties of preceding units, and depend on the
great convergence of inputs from level to level. None the less,
the apparent conceptual simplicity of the organization of the cat's
visual system is still open to considerable doubt as receptive
fields with increasingly complex properties are found at more
and more peripheral sites in the visual system. The two papers
suggested for further reading give examples of some more recent
experiments.

Further Reading

Stone, J. and Fabian, M. (1966). 'Specialised receptive fields of
the cat's retina.' *Science, N.Y.* **152**, 1277

Rodieck, R. W. (1967). 'Receptive fields in the cat retina: A new
type.' *Science, N.Y.* **157**, 90-93

Reproduced from *J. Neurophysiol.* **16** (1953). 37-68 by courtesy of
the Editor.

13

DISCHARGE PATTERNS AND FUNCTIONAL ORGANIZATION OF MAMMALIAN RETINA*

Stephen W. Kuffler

INTRODUCTION

The discharges carried in the optic nerve fibres contain all the information which the central nervous system receives from the retina. A correct interpretation of discharge patterns therefore constitutes an important step in the analysis of visual events. Further, investigations of nervous activity arising in the eye reveal many aspects of the functional organization of the neural elements within the retina itself.

Following studies of discharges in the optic nerve of the eel's eye by Adrian and Matthews [2,3], Hartline and his colleagues described the discharge pattern in the eye of the Limulus of a series of important and lucid papers (for a summary see Reference 20). In the Limulus the relationship between the stimulus to the primary receptor cell and the nerve discharges proved relatively simple, apparently because the connection between sense cell and nerve fibre was a direct one. Thus, when stimulation is confined to one receptor the discharge in a single Limulus nerve fibre will provide a good indication of excitatory events which take place as a result of photochemical processes. Discharges last for the duration of illumination and their frequency is a measure of stimulus strength. Lately, however, it was shown by Hartline et al.[22] that inhibitory interactions may be revealed when several receptors are excited. On the whole, the Limulus preparation shows many features which are similar to other simple sense organs, for instance, stretch receptors. In the latter, however, instead of photochemical events, stretch-deformation acts as the adequate stimulus on sensory terminals and is translated into a characteristic discharge pattern.

*This investigation was supported by a research grant from the National Institute of Health, U.S. Public Health Service.

The discharge from the cold-blooded vertebrate retina (mainly frogs) proved much more complex. Hartline found three main types when recording from single optic nerve fibres: (i) 'on' discharges, similar to those in the Limulus, firing for the duration of the light stimulus, (ii) 'off' discharges appearing when a light stimulus was withdrawn, and (iii) 'on-off' discharges, a combination of the former two, with activity confined mainly to onset and cessation of illumination. The mammalian discharge patterns were studied in a number of species by Granit and his co-workers in the course of their extensive work on the physiology of the visual system (summaries in References 13 and 15). On the whole, they did not observe any fundamental differences between frog and mammalian discharge types (see later).

The present studies were begun several years ago with the intention of examining the retinal organization and particularly processes of excitation and inhibition. As a first step, the discharge patterns were re-examined. It was assumed, in line with other workers, that the deviations in vertebrate eyes from the simple Limulus, or 'on' discharge type, are due to the nervous structures and to their interconnections between the rod and cone layers on the one hand and ganglion cells on the other. Therefore, an extension of such studies should shed further light on the functional organization of the retina.

A preparation was used which approached fairly satisfactorily the 'normal' state of the cat's eye. The discharge patterns reported by Hartline and those extensively studied by Granit were readily obtained. Single receptive fields—areas which must be illuminated to cause a ganglion cell to discharge—were explored with small spots of light and thereby some new aspects of retinal organization were detected. Specific receptive subdivisions, arranged in a characteristic fashion and connected to the common ganglion cells, seem to exist within each receptive field. This finding made it possible to study in detail some of the factors which normally contribute to the changing discharge pattern during vision. The present set-up also furnishes a relatively simple preparation in which the neural organization resembles the spinal cord and probably many higher centres of the nervous system. Many analogies have been found with discharge patterns in the spinal cord which are currently under study.

METHOD

The experimental arrangement, particularly the details of the optical system, has been described in full in a preceding paper[31]. The main instrument, the 'Multibeam Ophthalmoscope', consisted of a base which carried a holder in which the cat's head was rigidly fixed. Above the head, and also carried by the base, was the viewing-stimulating apparatus, which could be freely rotated and tilted. It contained three light sources with independent controls. This optical system was aligned with the cat's eye which thus was in the centre of a spherical co-ordinate system and the eye's ordinary channels were used for illumination of the retina. One light provided adjustable background illumination and thereby determined the level of light adaptation. It was also used as a source for observation of the retinal structures. A maximal visual magnification of about 40 was obtained. The background illumination covered a circle of not less than 4 mm (16 degrees for the cat) in diameter, centred on the recording electrode. Two Sylvania glow-modulator tubes were used for stimulation of restricted areas of the retina. They illuminated patterns, mostly circles of varying diameter, which were imaged on the retina. The smallest light spots were 0·1 mm in diameter on the retina. Thus, two images could be projected and their size and location varied independently on the retina. All three light sources used a common optical path, led into the eye through a pupil maximally dilated by Atropine or Neo-synepherine.

Complications from clouding of the cornea were prevented by the use of a glass contact lens, while the rest of the eye's optical system, lens and vitreous, remained intact. The circulation of the retina was under direct minute observation and whenever the general condition of the animal deteriorated this was readily noticed. The eye, as judged by its circulation and its discharge patterns, remained in good condition for the duration of the experiments which frequently lasted for 15-18 h. Dial-urethane (Ciba) anaesthesia (0·5 cm^3/kg) or decerebration was used. The effect of anaesthesia on the discharges is discussed later.

The eyeball was fixed by sutures to a ring which was part of the microelectrode manipulator. This fixation was generally satisfactory and breathing or minor body movement did not disturb the electrode position on ganglion cells. Sudden movements, however, such as coughing, jerking, etc., prevented continuous recording from single units. Occasionally a persistent slow nystagmus developed and, in order to abolish this movement, the tendons of extraocular muscles were severed at their insertions.

Microelectrodes were introduced into the eye protected by a short length of #9 hypodermic tubing which served to penetrate the scleral wall near the limbus. The unprotected electrode shaft, less than 1 mm thick, then traversed the vitreous and made contact with the retinal surface and toward the tip it was drawn to a fine taper. The shadow of the electrode thus covered only a small portion of the receptive field. If hit by the narrow light beams the electrode shaft caused scattering. All these phenomena and the positions and imagery of the stimulating beams or patterns were directly observed during the experiments and thereby a subjective evaluation of illumination conditions could be formed.

Electrical contact with the retinal cells was made by 10-15 μ Platinum-Iridium wires which were pushed to the tip of the glass tubes. The metal was either flush with the surrounding glass jacket which was sealed around it, or it protruded several micra. The configuration of the electrode tip was purposely varied a good deal, especially when the gan-

211

glion discharge was to be blocked by pressure. The potentials varied in size, and the largest were around 0.6 mV. The position of the indifferent electrode could be anywhere on the cat's body. In technically satisfactory preparations no difficulty was encountered in finding ganglion cells in quick succession and individual units could be observed for many hours (see later).

The second beam of the oscilloscope was used as an indicator of the current flow through the Sylvania glow-modulator tubes. The current was proportional to the light output but the spectral distribution of the light varied with different current strength. Therefore, Wratten neutral filters were used when the white light of the stimulators had to be attenuated. For the purpose of the present experiments the wave length variation which occurred played no significant role in those cases where intensities were varied by current flow adjustment (see *Figures* 7 III, *8, 10*). The accurate electronic control of the stimulating light sources made an adjustment of flash durations quick and convenient. The time base was also recorded on the second sweep by intensity modulation through a square-wave oscillator.

Illumination values are given in metre candles; the calibration was made for flux reaching the corneal surface above the pupil and calculated for the area which it covered on the retina. Losses within the eye's media are neglected. The maximal available background illumination was about 6,000 metre candles at the retina and could be attenuated to any desired extent. Since 1 mc at the retina corresponds to 10 mL external brightness (*See* Reference 31) the samples illustrated here were taken well within the photopic range. Discharge patterns were, however, also studied in the absence of background illumination. In most experiments the exploring spot's intensity was approximately 100 mc.

RESULTS

1. Some Characteristics of Single Unit Discharge

Differentiation between ganglion and axon potentials. As a recording electrode of 5-15 μ diameter at the tip made contact with the surface of the retina, a mass of potentials was usually recorded on illumination of the eye. Very light touch of the retinal surface rarely yielded differentiated single unit potentials. The latter could, however, be obtained with a slight further advance of the electrode, still without marked pressure against the tissue. Different degrees of 'touch' and pressure were easily differentiated under close direct observation (see Method). The most common and most easily recorded potential seen in the retina was a polyphasic spike, starting with an initial positive deflection, similar to that shown in *Figure 1(b)*. Such potentials are generally set up by a small spot of light at some distance from the recording electrode. From this observation it follows that conduction to the recording lead has taken place and that the

212

potential is derived from a nerve fibre. The polyphasic shape is typical of conducted potentials in a volume conductor. Similar potentials are familiar from recordings in other parts of the nervous system where microelectrodes are employed. The propagated potentials in nerve fibres could be used in the present

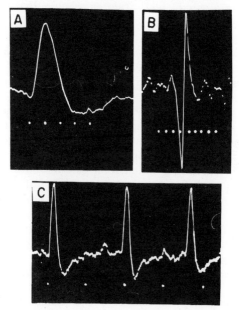

Figure 1. Potentials from different retinal elements recorded with microelectrode. *(a)* Ganglion cell discharge, caused by stimulation of retina in proximity of recording electrode. *(b)* Nerve impulse in an axon, set up by retinal stimulation some distance from electrode. *(c)* Three ganglion cell potentials from middle portion of a high-frequency discharge which is illustrated in *Figure 9c;* potentials become progressively smaller at this rate. Negative deflection of *(a)* and *(c)* 0·4 mV and 0·1 mV in *(b)*. Time intervals in *(a)* and *(b)* 0·1 ms and in *(c)* 1·0 ms. Note that ganglion potential can also start with small positive inflection if recording electrode is somewhat shifted.

studies, but they were small and could not be kept under the electrodes for prolonged periods. In contrast, potentials were recorded which always originated under the electrode tip *(Figure 1(a)*. These were simpler and larger and usually started with a sharp negative inflection which was followed by a relatively smaller positivity. The potentials were generally about 0·3–0·4 ms at the base, and their negative phase was of longer duration than in the potentials of *Figure 1(b)*, where the whole triphasic complex is of a similar duration. The distortion of the real potential time course is due to the smallness of the effective interelectrode distance with the present electrodes. In a volume conductor the potentials which arise close to or under the electrodes start with a sharp negative inflection, as in *Figure 1(a)*. On such grounds this potential is likely to be a ganglion cell potential which lies in the vicinity of the electrode contact. Physiological tests furnish convincing evidence for such a conclusion. The area of the retina, which on illumination caused discharges in the 'ganglion' cells, was found to be in the immediate neighbourhood of the electrode tip; this also was the place where the lowest intensity light spot was effective in setting up discharges. As an exploring spot was moved further from the tip of the electrode, stronger stimuli were needed. The active unit lay in the approximate centre of the 'receptive field' (*see* later) and excitation apparently reached it through converging pathways from its immediate neighbourhood. Such an arrangement is typical of ganglion cells. *Figure 1(c)* illustrates three ganglion potentials which form part of the high-frequency discharge series of *Figure 9c*; the impulses follow at intervals of 2·0 and 1·7 ms. At these rates the potential heights decline.

It follows from the relationship between receptive area and recording electrode that one can distinguish between conducted potentials in axons and those arising from ganglion cells. The latter may, however, also show a more complex shape, presumably when the recording electrode is some distance away from the cell body. The present technique favours the selection of larger ganglion cells but the extent of this selection is not known (*see* Discussion). The findings agree with those of Rushton[29], who by different methods showed the large single retinal discharges to arise from ganglion cells.

The potentials can also be easily distinguished by listening to their discharge in the loudspeaker. The ganglion potentials, which arise in the centre of the receptive fields, have a lower pitch, apparently because of less high-frequency components than in the axon spike.

Evidence for single cell discharges. The conventional criteria of single cell discharges are usually potentials of uniform size which arise at a sharp threshold and do not vary in a step-like fashion with fatigue or injury. Such criteria are generally sufficient to insure that potentials do not arise from several cells which fire in unison. In view of later findings, however, it is especially important to know that one really deals with single cell discharges.

The following procedures, which were incidental to many experiments, gave additional convincing evidence on this point.

(i) During *progressive pressure* which was obtained by advancing the recording electrode by means of a micrometer control, the ganglion cell discharge could frequently be blocked. Electrodes which had a relatively thick jacket flush with the

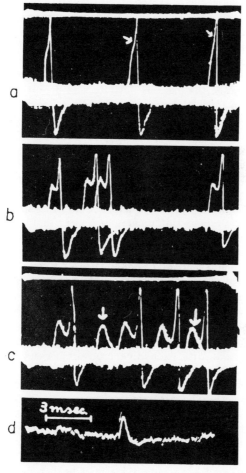

Figure 2. Progressive pressure block of ganglion cell discharge. Exposures made with sweep recurring at high frequency. Four successive stages of pressure block. *(a)* An initial inflection (arrow) appears on upper portion of rising phase. *(b)* A more discrete 'prepotential' is seen. *(c)* Prepotential is further reduced and occasionally (arrows) no spike appears. At second arrow a chance superposition of two apparently identical prepotentials occurred; one sets up spike, other fails to do so. *(d)* Spike is completely blocked, prepotential alone recorded on single sweep. Potential size in *(a)* is 0·3 mV.; note also that spike diminishes. Under progressive pressure potentials are of longer duration than normal (see Figure 1).

platinum tip, were most convenient for such pressure blocks. By these procedures, potentials could be separated into two components. The first component was variable over a very wide range and its height depended on the amount of pressure, while the other varied much less. The small potential had the characteristics of a local potential which precedes propagated spikes as described by Katz[25] and Hodgkin[23]. Accordingly, whenever such a 'prepotential' was sufficiently reduced the spike disappeared abruptly. These events leading up to pressure blocks are illustrated in *Figure 2*. Pressure itself frequently stimulated the ganglion cells and the ensuing activity was usually photographed by exposing a fast recurring sweep until a required number of impulses was obtained. In *Figure 2(a)* at the beginning of pressure, an inflection marked by arrows is seen on the upper half of the rising phase. In (*b*) the two phases are more marked, the spike taking off from the beginning of the falling phase of the prepotential. In (*c*) a critical level is reached and at the first arrow a pure prepotential appears. The second arrow indicates two potentials which, by chance, were accurately superimposed; in one case the prepotential causes a spike, in the other it just fails to do so. In (*d*) the prepotential alone is seen. It should be noted that the time course of the potentials under pressure is slower than under normal (*Figure 1*) conditions. This applies especially to the prepotential. While the microelectrodes give a distorted (shortened) time course of potentials, the difference between spike and prepotential seems significant. Decreasing the pressure restored the prepotential size and when it reached a critical height the spike suddenly reappeared; the process could then be repeated. With excessive pressure, however, the whole potential disappeared irreversibly. The constancy of the spike under such conditions of block, and recovery from block, confirm the assumption that it is derived from one ganglion cell only. It is unlikely that two cells should be so located in the vicinity of the electrode tip as to be affected in a quite similar and simultaneous manner. The origin of the variable prepotential was not studied in detail. It probably also originates in the ganglion cell, and such potentials may be set up there by the bipolars. It resembles some of the potentials obtained by Svaetichin[29] in spinal ganglion cells. Similar potential sequences are also seen at neuromuscular junctions or ganglionic synapses with curare or fatigue blocks.

(ii) The *potential size* of impulses at high frequencies is further evidence that single cell discharges are recorded. In the eye discharge, frequencies of 200-700/s and more are quite common. During these high-frequency bursts the potential size may decline, sometimes to about half of its original size. The decline is generally smooth in its progression and therefore cannot be due to one or two units dropping out during the discharge (*Figures* 9 and *10*). If one cell ceased to fire the potential should abruptly decrease. Alternatively it could hardly be assumed that several units should be so closely coupled. Variability of potential size in single peripheral nerve fibres has been observed at frequencies around 500/s when recording stretch receptor discharges[24]. There seem to be some differences, however, in the potential height changes between axons and ganglion cells. The latter tend to show a fall in height at lower frequencies, a fact already studied by Renshaw in spinal motoneurons[27]. In the present instances (e.g., *Figure* 9) the ganglion cell probably fires near its physiological limit, each impulse following in the relative refractory period of the preceding one.

The most convincing test of single unit discharge, however, was a functional one revealed in the mapping of the receptive fields. As will be shown below, discharge patterns are distributed in a characteristic fashion within receptive fields (e.g., *Figure* 6). That more than one ganglion cell should happen to have identical receptive fields with such a great regularity as was found in the present experiments would be a difficult assumption to make. Moreover, one would have to postulate that the cells always gave coupled high-frequency discharges at near-limit rates without, even occasionally, separating. Further, interaction, such as will be seen in the series of *Figure 8*, where regular mutual suppression of discharges occurs, could hardly happen if one recorded simultaneously from two or more cells.

2. *Spontaneous Retinal Activity*

Spontaneous activity in the mammalian retina has been regularly observed by Granit in dark-adapted cats[13]. In the present preparation considerable background discharge was a dominant feature especially in dimly illuminated retinae (1-5 mc at the retina). In dark-adapted eyes it proved very difficult to investigate the detailed discharge patterns of single units, since they fired frequently at 'resting' rates of about 20-30/s. The 'spontaneous' activity in the absence of illumination seems to be a normal feature for the following reasons: discharges due to injury of nerve fibres or ganglion cells under the recording lead, due to movement and pressure, could be excluded; spontaneous activity in many isolated units could be suppressed by illuminating the receptive fields some distance away from the recording lead (see also later); similarly, an electrode with a tip of 10-15 μ if gently placed near the middle of the optic disc, recorded massed spontaneous discharges which originated elsewhere, since illumination of the whole eye suppressed a great portion of the discharge; injury discharges along nerve fibres could not be expected to be modified by illumination in such a fashion.

Spontaneous activity was particularly pronounced in decerebrate animals, but was also regularly seen under Dial-urethane anaesthesia. The latter seemed to reduce the activity. Similarly, intravenous Nembutal, in amounts such as 20 per cent of the anaesthetic dose, had an immediate and prolonged effect in arresting or diminishing discharges from the retina. A similar effect with a slower onset was seen with intraperitoneal injections.

Since a great part of the present studies was done on cats under Dial-urethane the effect of the anaesthetic will influence the findings to an unknown degree. All essential observations, however, were also repeated in decerebrate preparations.

As indicated above, spontaneous activity, when recorded from isolated dark-adapted units, was generally suppressed or decreased for varying periods after application of increased background illumination. In the course of light adaptation, discharges usually returned gradually, or the slowed rates increased again. However, once a unit discharges in the light-adapted state, it is not possible to say how 'spontaneous' the activity is.

Of particular interest are those discharges which were apparently not due to injury and were not appreciably modified by general illumination of the eye. No detailed study of their nature could be made since they were never recorded in complete isolation. It is possible that during a steady increased background illumination many units appear which have previously not discharged, while others drop out. Such switching of active units may make it impossible to decide whether certain units have been continuously active or not. This important aspect of retinal activity has yet to be explored. In many cats grouped discharges in numerous nerve fibres were seen. They could usually be suppressed by illumination of the eye, but again their origin was not studied.

While most features of 'spontaneous' activity remain to be investigated, it is a noteworthy phenomenon, since it is upon such a high level of background activity that patterns of many

218

visual events are superimposed. Rhythmic and 'spontaneous' activity is common to the central nervous system in mammals and has also been observed in a variety of other visual systems[1, 4, 7].

3. Extent of Receptive Fields of Cat's Retina

The receptive field of a single unit was defined by Hartline as the area of the retina which must receive illumination in order to cause a discharge in a particular ganglion cell or nerve fibre. Hartline[17, 18] was the first to study the physiological characteristics of receptive fields of single optic nerve fibres in frogs in a precise and thorough manner, by exploring the area with a small spot of light. Since the retina is composed of a group of overlapping receptive fields, the extent of these is of obvious interest. By charting the boundaries of an area over which a spot of light sets up impulses in a ganglion cell or in its nerve fibre, one will obtain the configuration of the receptive field. The field size depends on stimulus strength, the size of the exploring spot and the state of dark adaptation. The latter will largely determine the level of sensitivity of the area. For instance, if an exploring spot is made smaller, or if the level of background illumination is increased, the intensity of the spot has also to be increased in order to set up responses over as large an area as previously. The problems of determining receptive field sizes have been discussed in detail by Hartline[18], and his results on frogs were found to apply equally to the mammalian retina.

The receptive field definition may be enlarged to include all areas in *functional* connection with a ganglion cell. In this respect only can the field size change. The anatomical configuration of a receptive field—all the receptors actually connected to a ganglion cell by some nervous pathways—is, of course, assumed to be fixed. As will be seen below, not only the areas from which responses can actually be set up by retinal illumination may be included in a definition of the receptive field but also all areas which show a functional connection, by an inhibitory or excitatory effect on a ganglion cell. This may well involve areas which are somewhat remote from a ganglion cell and by themselves do not set up discharges.

The optical conditions in the mammal present additional difficulties for mapping of receptive fields, as contrasted to those in the opened frog's eye. Because of the imperfections of the optical system, an appreciable amount of light scattering occurs and the images will be less sharply focused. The most advantageous situation for the full exploration of the receptive fields, which approximates the anatomical receptive field boundaries, is complete dark adaptation. During this state, however, most units discharge spontaneously, making threshold determination or detection of changes in response patterns difficult. The mapping was mostly carried out in different states of light adaptation, and even under such conditions a 'steady' state cannot be maintained. As implied in the term 'adaptation', thresholds change, drifting towards a lower value, and discharge patterns may also vary correspondingly. Such changes seem to be part of normal events in the eye. In spite of these factors some relevant data of the size of receptive fields can be obtained.

Figure 3 illustrates a chart of a retinal region which contains receptors with connections which converge upon one ganglion cell and cause it to discharge. The exploring spot was 0·2 mm

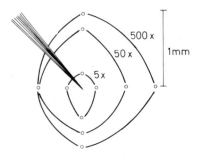

Figure 3. Extent of receptive field obtained with exploring beam of 0·2 mm in diameter at three different intensities. Electrode (shaded) on ganglion cell. Background illumination about 10 mc. Inner line encloses retinal region within which light spot, about 5× threshold at electrode tip, sets up discharges. Other boundaries of field were mapped at intensity 50× and 500× threshold. Note that on left, receptive field does not expand appreciably as stimulating spot intensity is increased.

in diameter and the background illumination approximately 10 mc. The smallest inner area was obtained by an exploring spot, about five times threshold for a position near the electrode tip. If the spot was moved outside this area (5×), no discharges were set up. If the spot intensity was increased 10 times, by removing a Wratten neutral filter, and thus making it 50 times threshold, it caused discharges within the larger area (50×). Further increase in the stimulus strength to 500 times threshold expanded the receptive field on three sides (500×) while the demarcation line on the left remained practically unchanged. This may indicate that light scattering was not a very great factor in this particular mapping. Otherwise such a fixed portion of the boundary, in spite of an increase in stimulus intensity, could hardly be obtained. Frequently a receptive field as shown here was charted and then the exploring spot was further increased in strength. The field suddenly expanded several times and then generally no distinct boundary demarcation was obtained. It is thought that this was clearly due to scatter of light since a reduction of the stimulating spot size again resulted in a definite limit of the receptive field.

The present technique, using small exploring spots, is suited to detect relatively dense concentrations of receptors which feed into a single ganglion cell, and therefore provides only an approximate estimate of the actual anatomical receptor distribution. Evidence suggests that the density of receptors beyond the receptive field limit *(Figure 3)* may be insufficient to produce more than subthreshold effects on a ganglion cell (see Discussion). Stimulation with larger spots may overcome the difficulties and extend the receptive field into areas where the receptor concentration is low. By increasing the spot size, in fact, receptive fields apparently 3-4 mm in diameter were found, but scatter of light makes those findings unreliable. The experiment should be done by the use of a great variety of illumination patterns near threshold intensities which would allow a more exclusive excitation of the 'surround', while the central region is not illuminated. Most determinations in the present study were made in the region of the cat's tapetum, a highly reflecting region where the anatomical features of the retina can be observed with greater accuracy through the optical system. Further, the tip of the recording electrode can be seen, the stimulating spot can be followed, and in this way conditions can be checked by direct observation, provided the background illumination is sufficiently bright. The receptive field diameters varied between 0·8 and 2·0 mm with the present method. Small ganglion cells may have fields of different extent. No determinations have been made in the periphery of the retina (see Discussion).

4. Stimulation of Subdivisions of Receptive Fields

(a) *Specific areas within receptive fields.* In Hartline's[17] experiments stimulation anywhere within a receptive field of the frog caused essentially the same discharge pattern in a given fibre; i.e., either 'on,' 'on-off' or pure 'off' responses resulted. Accordingly the discharge type from the frog's receptive field seems relatively fixed (see, however, Discussion). This question was investigated in the present study.

It was found that the discharge patterns from ganglion cells whose receptive fields were explored varied with the specific subdivisions which were illuminated. *Figure 4* illustrates such findings. A light spot, 0·2 mm in diameter, was moved to different positions, all within an area of 1 mm in diameter. In *Figure 4 (a)* a discharge appeared during illumination; this 'on' response was of a transient nature and although stimulation was continued at the same intensity, the discharge ceased within less than one second (*see* Section 6). In *Figure 4(b)* when the light spot was moved 0·5 mm from the first position no 'on' discharge at all appeared and the response was of the pure 'off' type, i.e., discharges occurred after the cessation of illumination. At an inter-

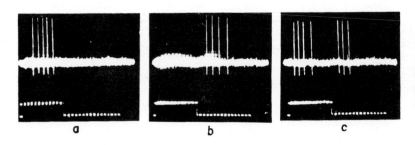

a   b   c

Figure 4. Specific regions within receptive field. 0·2 mm diameter light spot moved to three different positions within receptive field. Light flash to region near electrode tip in *(a)* causes only 'on' discharges in ganglion cell, while same stimulus 0·5 mm away is followed by 'off' responses *(b)* and in an intermediate position an 'on-off' discharge is set up *(c)*. In this and subsequent records second beam signals intensity and duration of light flash; intensity modulation of 50/s gives time base. Impulses 0·5 mV

mediary position of the exploring spot, a combination of the first two responses resulted, and an 'on-off' discharge is seen (c). All transitions in discharge patterns from those here shown were seen when the light spot of fixed intensity was moved to a number of positions within the receptive field, while the background illumination of the eye remained constant. Other illustrations of changes in discharge patterns with illumination of different areas within the receptive field are seen in *Figures 7* and *8*. Thus, the ratio and number of 'on' or 'off' discharges varied with the specific area which was illuminated. The changes in discharge type, caused by merely shifting an exploring spot, were not always striking in all units. To obtain the varied discharge patterns it was frequently necessary to change, in addition, the state of light adaptation, the stimulus intensity, or area of the stimulating light (see below).

It is concluded that within the receptive fields of single gang-lion cells (or nerve fibres) there exist areas which can contribute differing discharge patterns. The discharge, as seen with stimu-lation of the whole receptive field, is the resultant of the con-tribution and interaction of all of these areas.

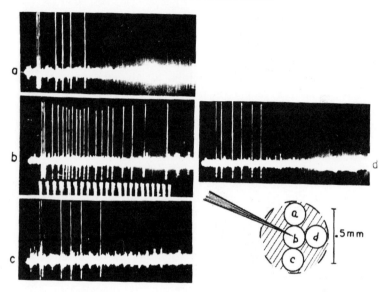

Figure 5. Centre portion of receptive field. Ganglion cell activity caused by circular light spot 0·2 mm in diameter, 3-5 times threshold. Background illumination was about 30 mc. Positions of light spot indicated in diagram. In *(b)* an 'on' discharge persists for duration of flash. Intensity modulation at 20/s. Movement of spot to positions *(a)*, *(c)*, and *(d)* causes lower frequency discharge which is not maintained for duration of light stimulus. Movement of spot beyond shaded area fails to set up impulses (see, however, extent of receptive field in similar unit with stronger stimuli in *Figure 6*). Potentials 0·5 mV

(b) Distribution of discharge patterns in receptive fields. All units had a central area of greatest sensitivity in which either the 'on' or the 'off' component predominated in the discharge pattern. Flashes of 0·5-1·0 s duration, for instance, to sub-divisions of an area of perhaps 0·5 mm in diameter around the ganglion cell would give 'on' responses only. Within this area the 'on' frequency decreased as the spot was shifted away from the most sensitive region in the centre. This is shown in *Figure 5*. A spot 0·1 mm in radius was projected onto the retinal region

around the tip of the recording electrode which was placed on a ganglion cell. In this and nearly all other experiments the region of electrode contact proved to be the most light-sensitive part of the receptive field. The area of lowest threshold and the geographical centre of the receptive fields usually coincide. If the stimulating light spot was made 3-4 times threshold for the central location it evoked there a vigorous 'on' response for the duration of illumination (*Figure 5(b)*). A shift of the light spot, as illustrated in the scheme included in *Figure 5*, made it much less effective. The 'on' discharges set up by the same stimulus became shorter and of lower frequency, and with further movement away from the centre no discharges at all were set up. The boundaries of the receptive field with this relatively weak stimulus strength at a background of 30 mc are indicated by the broken circle.

The records of *Figure 5* show only a central area of a receptive field similar to the one which is within the inner circle of *Figure 3*. If the small exploring spot is made 100-1000 times threshold, a more complete picture of the discharge pattern distribution in receptive fields can be formed. The chart of *Figure 6* was obtained from a unit under a background illumination of about 25 mc. It is characteristic in a general way of the majority of units which have been studied. The crosses denote 'on,' the open circles 'off' responses, and the 'on-off' discharges are indicated by the cross-circle combinations. The different shaded areas give an approximate picture of the predominant areal organization within the receptive field, i.e., of receptors and neural connections (see Discussion). The centre-surround relationship may be the converse in other units, with the 'off' responses predominating in the centre; the area ratio between centre and surround also fluctuates greatly. Further, the discharge pattern distribution shifts with changing conditions of illumination (see below).

Not in all units was the field laid out in a regular concentric manner as in *Figure 6*. The areas were frequently irregular. In some instances there appeared 'gaps' between regions; i.e., isolated spots in the periphery seemed to be functionally connected to a ganglion cell.

(c) *Factors modifying discharge patterns and size of receptive fields.* As indicated above, the discharge patterns arising in single receptive fields may vary, if conditions of illumination are altered. The four upper records of *Figure 7* show 'on' dis-

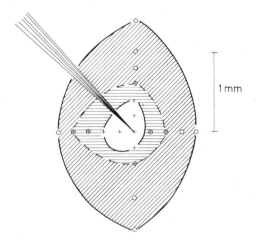

Figure 6. Distribution of discharge patterns within receptive field of ganglion cell (located at tip of electrode). Exploring spot was 0·2 mm in diameter, about 100 times threshold at centre of field. Background illumination approximately 25 mc. In central region (crosses) 'on' discharges were found, while in diagonally hatched part only 'off' discharges occurred (circles). In intermediary zone (horizontally hatched) discharges were 'on-off'. Note that change in conditions of illumination (background etc.) also altered discharge pattern distribution *(see text)*

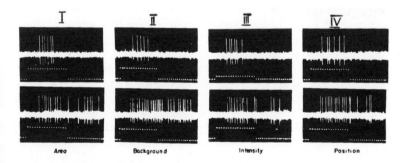

Figure 7. Change in discharge pattern from 'on' response (upper records) in single ganglion discharge into an 'on-off' response (lower records). In I stimulating spot of 0·2 mm diameter in central region of receptive field set up 'on' discharge. Increasing spot diameter to 3 mm set up more 'on' impulses and brought in an 'off' component. Same result was obtained in II by merely decreasing background illumination from 19 mc to 4 mc and in III by increasing stimulus spot intensity (intensity scale in III different). In IV exploring spot was shifted by about 0·4 mm from central into more peripheral part of receptive field. Intensity modulation 50 p.s.

225

charges produced by a 0·2 mm diameter light spot. In the lower records is seen a corresponding series of 'on-off' discharges which were obtained from the same unit by changing different parameters of illumination. In I the area of the stimulating spot was increased so as to include the whole receptive field and thereby the 'on' was converted into an 'on-off' discharge. In II the same effect was obtained by decreasing the background illumination while leaving all other conditions unchanged. In III merely the intensity of the testing spot was increased, while in IV the spot was moved to another portion of the receptive field, without altering its intensity or area. It follows from these observations that a modification of any of these variables of light stimulation, alone or in combination, will in turn lead to modifications of the discharge pattern. In addition to the factors illustrated in *Figure 7*, the duration of stimulation also plays a role. The direction of the changes can usually be predicted. If, in a composite discharge pattern, one of the components—for instance, the 'on' portion—predominates strongly, a reduction of stimulus strength will cause the relatively weak 'off' fraction to disappear, while the 'on' may be only little affected. The same result can generally be obtained by merely increasing the background illumination or reducing the area of the stimulating spot. Conversely, a combination of a weak 'on' and a strong high-frequency 'off' component can be changed into a pure 'off' response by reducing the stimulus strength or increasing the background illumination intensity. Discharge patterns can frequently be altered by variation of background and stimulating light intensities even when the whole receptive field is illuminated. However, results are usually not as clear-cut as with fractional activation of the receptive field.

The *effect of background illumination* deserves more detailed analysis since it is one of the most potent factors in altering discharge conditions. As the background illumination is increased, the boundaries of the receptive fields 'contract' and also the discharge pattern distributions change. The response type which is characteristic of the surround (diagonally hatched area of *Figure 6*) tends to disappear and the pattern of the centre (non-hatched region) will predominate. In fact, some units even with careful exploration, using small 0·1-0·2 mm light spots under photopic conditions, gave only pure 'on' or 'off' responses within the limits of the receptive field which might be only 0·5 mm in diameter. If the area of the stimulating spot was increased without changing its intensity—for instance, by illuminating a retinal patch 1 mm in diameter—then the stimulus occasionally brought in an additional weak response which was characteristic of the 'fringe' or surround. Thus, an 'on' type of response would be converted into an 'on-off' as the spot size was increased (see also *Figure* 71). The characteristic response of the surround could always be made evident by using a dim background, or after a short period (several minutes) of complete dark adaptation. Decreasing the background illumination first expanded the area from which centre-type responses could be elicited, then brought in 'on-off' responses around its boundary and eventually disclosed discharges which were characteristic of the surround. Whenever a careful search was made, both 'on' and 'off' components were seen in all receptive fields.

It should be noted that increased background illumination changed the receptive field in a similar manner in all units which were studied. The surround type of response, involving a presumably less dense contribution of receptors (see Discussion), was always suppressed first, independently of whether it consisted of a predominantly 'on' or 'off' response. This will have to be considered in discussions of the contribution of rods and cones to discharge patterns.

The great range of flexibility at the level of the single unit discharge is of particular interest, since all the factors which were found to affect the discharges play a role under normal conditions of vision.

5. Interaction of Different Areas within Receptive Field

It may be presumed that one of the basic contributions of interneurons within the retina (cells between the photoreceptors and the ganglion cells which give rise to the optic nerve fibres) consists in modifying the pattern of discharges which are set up by excitation of rods and cones. The impulses emerging through the optic nerve show the result of a complex series of events which have taken place in the retina, such as spatial interaction and processes of facilitation and inhibition. These problems have already been considered by Adrian and Matthews in their classical investigations on the eel's eye[2, 3] and by Hartline in the early studies of the organization of the receptive field[17, 18, 19]. A wealth of data on the functional organization of the retina has also emerged from Granit's laboratory[13, 14, 15].

An additional approach is made possible by the present findings that certain areas within a single receptive field make a

suppressed (*Figure* 8 I, A + B.). At the same time the number of 'on' impulses was somewhat reduced as compared with the control response to stimulation of spot (A). Such situations could be produced regularly with two spatially separated light spots within a receptive field, i.e., illumination of one area could suppress discharges arising from stimulation of another. The reverse situation from *Figure* 8 I could be produced in the same unit as is shown in *Figure* 8 II. Spot (A) was made less intense while (B) in the surround was made stronger. When these stimuli were given together (A + B), the 'on' effect of (A) was completely suppressed, while the off discharge was but little affected. An intermediate situation between *Figure* 8 I and II could also be created by altering the intensities so as to make the effects from spots (A) and (B) equally 'strong'. When flashed simultaneously in *Figure* 8 III (A + B), they simply reduced each other's effect, setting up a relatively weak 'on-off' response. In order to make certain that increased scatter of light with two spots was not responsible for the effects, the two light beams were super-imposed. In such cases their effect on the discharge was simply additive. *Figure* 8 illustrates only a few of the possible variants in discharge which can be produced by two interacting light patches. Instead of changing the intensities of the stimulating. spots, results similar to *Figure* 8 could also be obtained by merely varying the areas of spots (A) and (B) so as to produce the required amount of 'on' or 'off' discharge. Alternately, shifting the location of the light stimuli or altering the background illumination would balance the 'on' and 'off' relationship in any required direction.

In many experiments one light spot was fixed and the other was moved around it in the manner of a satellite. In this way a systematic study was made of the interacting regions within a receptive field. As might be expected from the above results, one could produce all combinations of response types and variants of the 'on-off' ratio. Once the receptive field with its boundaries and discharge patterns within that area was plotted (see *Figure* 6), the result of interaction of two spots could usually be predicted. It is worthy of note that in the present experiments not only the excitatory result of a light stimulus, such as an 'on' discharge, could be inhibited, but also the 'off' discharge—itself a consequence of inhibitory processes—could be suppressed. As a rule, then, when two light stimuli within the receptive field interact, *both* become modified, but if the effect of one is much 'stronger' than the other, its discharge may not be appreciably affected.

Suppression of 'off' responses could also be seen some time

after stimulation of an 'on' area. The time course of this inhibitory effect, presumably caused by persistent excitation after previous illumination, could be studied in the following manner. In units similar to that shown in *Figure 8*I the duration of the stimulus to the 'on' area (A) could be progressively shortened while (B) was kept constant. It was found that beam (A) could suppress (B) for varying periods after (A) had been turned off. The time course of the inhibitory after-effect depended on the duration and intensity of (A). There was a transition from complete suppression of the 'off' discharge to partial suppression and to a mere delay in the onset of the 'off' discharge.

In these investigations it was surprising that frequently a ganglion cell, which gave an 'off' effect, was largely unresponsive to stimulation of an 'on' area during the period of the 'off' discharge. Further, in the tests where the interaction of two 'on' areas was studied, lack of addition of excitatory influences frequently developed. Since these observations on interaction phenomena have a bearing on functional organization of the retina a more thorough analysis will be presented in a separate publication.* Particularly the combination of spatial and temporal effects opens up some further approaches. These instances are mentioned here because they present a wider picture of factors which play a role in the production of discharge patterns. Further, they tend to explain some 'anomalous' observations, such as lengthening of latent periods with stronger stimuli, or increased discharge frequencies with weaker ones (*Figures 11* and *12*).

6. Characteristics of 'On' Discharge

(a) *Transient and maintained 'on' response.* From analogies with the Limulus eye there may be reason to suspect that the maintained 'on' response in mammals, which keeps discharging for long periods during illumination, is set up in receptors which have a fairly 'direct' connection from photoreceptors to bipolars and to ganglion cells. On the other hand, the 'on' which is part of the frequently occurring 'on-off' type may be set up in units where the receptive field has different neuroanatomical connections.

In the preparations studied there were units which gave only the Limulus type of 'on' response when the whole retina was stimulated under photopic or scotopic conditions. Under careful

---

*These questions are discussed more fully in *Cold Spr. Harb. Symp. quant. Biol.*, (1952) *17*.

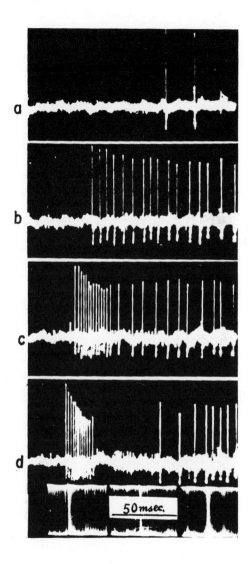

Figure 9. Effect of stimulus strength on latent period and discharge frequency. 0·2 mm diameter spot projected onto 'on' centre of a receptive field at illumination background of about 2 mc. Between *(a)* and *(d)* stimulus was increased in steps of 10. Latent periods were 93, 36, 22 and 15 ms. Peak frequency in *(d)* was over 800/s. Transient pause after a high-frequency burst occurred regularly, as did decline of potential size. Impulses 0·4 mV

231

scrutiny, when restricted subdivisions of the receptive field were stimulated, with dim background illumination, it was always observed that these 'on' units also received 'off' contributions from the periphery (see Section 4). More frequent were those units which gave a transient 'on' response lasting about 1-2 s

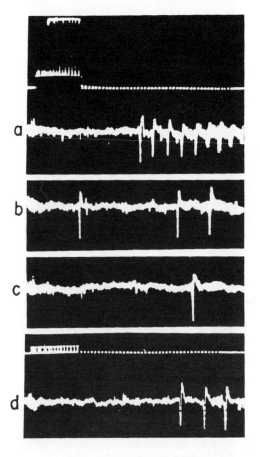

Figure 10. Ganglion discharge with spot (0·2 mm diam.) illumination *(a)* flash of 6·5 ms in duration to centre of receptive field set up response with initial frequency of 575/s and latent period of 15 ms. Prolonging illumination did not change latent period but caused an 'on' response for 2-3 s; *(b)* same flash, image moved 0·1-0·2 mm from central position. Latent period 21 ms, only two impulses set up. First impulse on sweep was 'spontaneous' and not related to flash. *(c)* same flash as *(a)* but background illumination increased. Only one impulse set up. *(d)* conditions as in *(a)* but stimulus intensity decreased. Latent period 22 ms, discharge burst shorter. Effects seen in *(b)-(d)* were also obtained by shortening flash or reducing spot size. Intensity modulation 2,000/s.

with diffuse maintained retinal stimulation. These were generally followed by an 'off' response, depending on the background illumination (*see* above). The most frequent units were those with 'on-off' discharges, the 'on' lasting 1 s or less. The following modifications of the transient 'on' responses were of special interest because they revealed some further aspects of receptive field organization: (i) when the central portion of some receptive fields was illuminated by a spot of 0·1-0·2 mm in diameter an 'on' discharge resulted lasting for seconds or, in several instances, even minutes. Either increasing or decreasing the stimulus strength frequently shortened the duration of the 'on' discharge. (ii) Moving the stimulating spot as little as 0·1-0·2 mm from the centre of the receptive field greatly shortened the discharge and at the same time the onset of the discharges could be delayed (*Figure 10(b)*). Further, units were observed which gave a maintained 'on' response at the centre, 'off' responses in the periphery and transient 'on' responses coupled with 'off' discharges in intermediate regions of the receptive field. This required the selection of an appropriate background illumination, stimulating intensity and size of the exploring spot. (iii) In some units a small central spot gave maintained 'on' responses and, as the area of the illuminating patch was enlarged to include the surround, the discharge became of the transient type (*see also* Reference 17). (iv) One isolated instance in which, however, the unit gave easily repeatable responses for several hours deserves mentioning. Under a background illumination of 10-20 mc the unit showed an 'on' response which could not be maintained for longer than 1-2 s at any available intensity of the stimulating spot which was 0·2 mm in diameter and directed onto the central region. When the background illumination was increased (60-100 mc) this discharge was converted into a maintained 'on' type although the stimulus was of the *same* intensity as that which gave the shorter 'on' response before. This situation was the reverse of the more common one since, by increasing the background illumination, a given stimulating intensity usually becomes less effective. One may surmise that in this unit the background illumination preferentially suppressed inhibitory influences from the 'off' areas.

The above findings suggest the following interpretation: the maintained 'on' discharge is converted into the transient type by activation of elements which converge onto the same ganglion cell from the periphery of the receptive field. Accordingly a unit which is so organized that it has a strong 'on' centre and a weak 'off' surround will tend to give a well-maintained discharge even with illumination of the whole eye. The discharge will shorten in proportion to the peripheral 'off' contribution. Such a view is also supported by the interaction experiments in which a simultaneous second spot in the surround weakens and

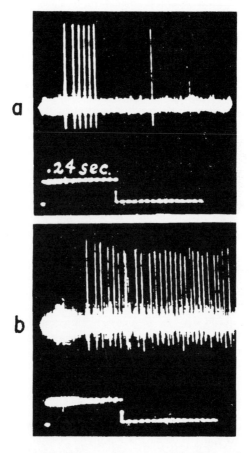

Figure 11. 'Anomalous' effect of change in stimulus area on latent period. *(a)* ganglion discharge set up by 0·2 mm diam. spot within central region of receptive field. *(b)* spot size increased so as to include whole field. Note the greatly prolonged latent period of 'on' component. Potentials 0·6 mV

shortens the discharge set up by the central one. The duration of the 'on' discharge then will depend on how many 'off' pathways to a ganglion cell are active in relation to the 'on' fraction.

234

It is realized that the inhibitory 'off' action starts approximately simultaneously with the 'on' action. Therefore, if both continued simultaneously at the same strength, one would expect merely a reduction of the 'on' discharge frequency scale and not a shortening when a certain 'off' component is added. Such a reduction of an 'on' discharge is seen in *Figure 8* III. However, the 'on' discharges which are generally observed start at a relatively high frequency which subsequently tends to decrease. With

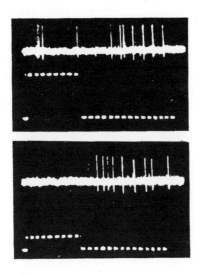

Figure 12. 'Anomalous' effect of change in stimulus intensity on discharge. Upper record: 'on-off' ganglion discharge. Below: with stimulus intensity decrease 'on' component drops out. Note, however, the shorter latent period and increased number of impulses in 'off' discharge (*see* text). Frequency 50/s.

reduction of the stimulus strength producing such an 'on' discharge, the initial high frequency will be reduced while the later discharge of lower frequency may drop out completely (*see also* Reference 18). Therefore a similar 'weakening' of an 'on' discharge by an inhibitory action may lead to a shortened 'on' response. Further, the suppressing effect from 'off' zones does not necessarily start with its full force, but may increase with prolonged stimulation as can frequently be seen in its action of stopping 'off' discharges. The presence of inhibitory contributions in many pure 'on' elements has already been shown by Donner and Willmer[9].

235

(b) *Latent period and discharge frequency of 'on' responses.*
Generally one can cause increased excitation, as measured by
frequency of response and shortened latent period, by (i) increased stimulus intensity, (ii) increase in stimulated area, (iii)
decrease in background illumination (or increased dark adaptation), (iv) moving the stimulating spot toward the centre of the
receptive field.

A fairly typical effect of stimulus strength on the latent period
and discharge frequency is seen in *Figure 9*. A spot 0·2 mm in
diameter was flashed at four different intensities onto the 'on'
centre of a receptive field, increasing in steps of 10 from (a) to
(d), with the eye under a background illumination of about 2 mc.
This illustration is of particular interest because it shows how
short the latent period can be and how high the discharge rate
can become in the cat's retina even with moderate intensities of
stimulation. In (a) the stimulus is near threshold and the latent
period is 93 ms. In (b) the latency is 36 ms and the average
discharge rate for the first 8 impulses is about 180/s. In (c) the
discharge frequency is 300/s for the first 13 discharges and the
latency is 22 ms and in (d) a peak frequency of over 800/s is
reached between the fourth and tenth impulse, the latency being
15 ms. This rate of discharge is much higher than is customarily
obtained from nervous structures under physiological conditions.
A pause as in (d) is common, both after 'on' or 'off' bursts.
Increasing the area of stimulation within the centre of the receptive field, starting with a relatively weak stimulus, also caused
higher discharge frequency and latency shortening in this unit
(see, however, below). The latent period of 15 ms in *Figure 9(d)*
is shorter than hitherto seen in mammals, presumably due to
restriction of the stimulus to a predominantly 'on' area (see
below).

Figure 10 illustrates a unit which gave an 'on' discharge lasting several seconds with illumination of the whole eye and a
somewhat longer one with illumination confined to its 'on' centre.
In (a) it showed a high-frequency initial burst of 575/s for the
first 8 impulses with the potential size sharply declining (followed by a pause). The latent period of 15 ms in (a) was lengthened
and the discharge frequency and duration were reduced in the
subsequent three records by the following: (i) in (b) the light spot
of the same intensity as in (a) was moved from the centre of the
field by 0·1 to 0·2 mm; (ii) in (c) with the light spot in the centre
again, the background illumination was increased; and (iii) in
(d) the stimulus intensity was reduced. Reducing the stimulating

area or shortening the duration of the light flash (not illustrated) had a similar result as shown in (b)-(d). The findings of *Figures 9* and *10* are in general agreement with the early work of Adrian and Matthews[2], Hartline[17] and Granit[13].

Some notable exceptions to the general 'rules' as discussed above were also observed—and, in fact, could frequently be produced by appropriately arranging the conditions of the experiment. Thus, in contrast to the usual results, the latent period of discharge was actually prolonged in the unit of *Figure 11* when the area of stimulation on the retina was changed from a patch 0.2 mm in diameter (a) to one of 3 mm (b). Similarly, increasing the light intensity could have the same effect. One may assume that stimulation of the larger area brought in a strong 'off' component from the surround, causing a delay in the 'on' response. Such a situation could actually be produced frequently by stimulation of two separate small 'off' and 'on' areas. Another 'exception' is seen in *Figure 12* where an 'on-off' response is converted into an 'off' by reducing the stimulus strength. Surprisingly, however, the latent period of the 'off' response is shorter and the number of impulses is greater with this weaker illumination (*see also* Reference 9). Again, an explanation can be sought in the antagonism of 'on' and 'off' influences. The weaker stimulus, by failing to excite the 'on' fraction, caused less inhibition of the 'off' component. In all these 'anomalous' instances it must be noted that a non-homogeneous population of receptors is activated and the discharge pattern depends on the proportion of 'off'- and of 'on'-oriented receptors which are excited.

7. 'Off' response

As appears from Section 4, no pure 'off' units were found when the receptive fields were explored with small spots of light and suitable background illumination. Those units which gave an 'off' response alone with illumination of the whole eye were always found to have an 'off' centre and 'on' surround, while units giving 'on-off' responses could have either type of centre. The 'off' activity of an area could be tested by the ability of a light stimulus to set up impulses when its intensity was reduced or the light turned off, or by the suppression of spontaneous activity.

The interaction between separate stimuli to 'on' and 'off' areas was shown in *Figure 8*; in *Figure 13* a similar experiment is illustrated with both light stimuli to an 'off' region. Spot A caused a strong 'off' response by illumination of an area 0·2 mm in diameter in the central portion of a receptive field, just about 0·1 mm away from the area of lowest threshold at the electrode tip. The illumination was started before the sweep and only the cessation of the light signal appears on the record (marked by arrow).

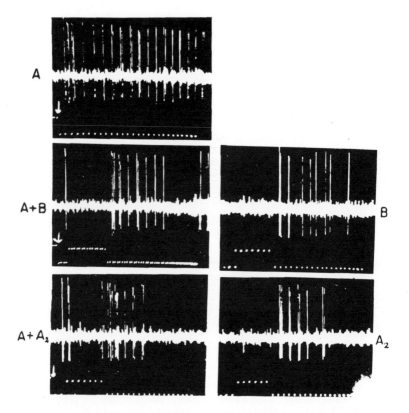

Figure 13. Inhibitory action of light on 'off' response. Light beams (A) and (B) projected onto separate areas, each 0·2 mm in diameter, in central region of a receptive field near tip of recording electrode. Both regions give 'off' responses only. Background 18 mc (A) 'Off' discharge produced following termination (arrow), at beginning of sweep, of stimulation by beam A. A+B: Beam B, applied during 'off' discharge, suppresses impulses. B: spot B alone. A+A₂: Stimulus to spot A ceases near beginning of sweep, as above, but same area is re-illuminated by second flash. Not only is there suppression of 'off' discharges during flash of A₂, but also subsequent 'off' response duration is reduced as compared with A. A₂: second flash of beam A alone. Note that 'off' discharges set up in one region of receptive field can be suppressed by stimulation of another 'off' region, or by restimulation of same area. The grouped discharges occurred in many units of this experiment. Time base 100/s in A+B 50/s in all other records Potentials 0·3 mV

238

Grouped discharges similar to those in this figure were frequently seen and have been also noted by others during the 'off' effect[16, 17]. Spot B was the same distance from the electrode tip as A, but on the opposite side. This stimulus was shorter and by itself caused a briefer discharge (*Figure 13B*). When B followed A, it suppressed the 'off' discharge for the duration of its flash (*Figure 13A + B*). When both stimuli were given to spot A in succession, the second (A_2) also suppressed the 'off' impulses. A_2, being a shorter flash than the preceding A, set up a shorter 'off' response than A alone. It is noteworthy, however, that the 'off' discharges of A were not reinforced at the end of flash A_2. In this unit it seems that A_2 during its flash not only suppressed the impulses, but also the processes which 'survived' after cessation of A. The inhibitory action of light on the 'off' discharge by re-illumination of the whole eye is well known from the work of Granit and Therman[16] and Hartline[17]. Suppression of 'off' discharges, set up in one region of the receptive field, by subsequent excitation of another 'off' area is to be expected from the experiments on interaction (Section 5) and has also been seen by Hartline[20a].

The duration of the latent period of the 'off' responses was studied, since it is a measure of the processes which have preceded the discharge. It is known that the latent period shortens and the discharge frequency increases as a function of the intensity and duration of the preceding illumination[13, 20]. Again, however, exceptions to this rule occur. In some experiments latent periods as short as 10-15 ms, similar to the shortest periods for 'on' discharges, could be seen. Another indication concerning the processes which are involved in inhibitory activity can be obtained from a determination of the time which is taken up between stimulation of the receptors and the first sign of suppression of activity at the ganglion cell level. Some conclusion may then be drawn regarding the mechanism of excitation spread within the retina. The speed and mode of this spread will be important in the competitive situation when both 'on' and 'off' areas are excited simultaneously, as must occur normally in the eye when stimulation is not confined to subdivisions of a receptive field. Such latent periods of inhibitory action are obtained by measuring the time it takes for a second flash (B or A_2 in *Figure 13*) to suppress a discharge. The time between the onset of the 'off' flash and the first suppressed impulse would clearly be the most accurate determination. This method will be most precise if the suppression is tested and measured on a well-maintained and regular high-frequency discharge. By such determinations the shortest latent periods of inhibition were around 10 ms. These times may, in fact, be too long since they do not indicate the actual onset of inhibitory action at the ganglion cell. The processes may start acting well before they become evident by their action of suppressing a discharge. Further information in this connection will be presented in a study of the inhibitory and excitatory pathways which converge on ganglion cells.

DISCUSSION

Sampling of units within retina. An advantage of the present technique is the ease of recording retinal activity and the intactness of the eyeball which enables the normal optical channels to be used for illumination and observation. The method, however, will tend to select the potentials from the larger ganglion cells. On the other hand, since one generally can find suitable cells for recording within any small area of the retina, such as 1 square mm, it is quite likely that these cells can be smaller than the 'giant' cells described by Rushton[27]. Further, nearly all the work was done on cells within a radius of about 5-8 mm from the optic disc, particularly in the two quadrants above the disc within the highly reflecting region of the tapetum. No positive evidence has been found that within these areas there are specific subsections which give different discharge patterns. The cat has no fovea but there exists a region on the visual axis of the eye, called centralis[6] about 1-2 mm temporal from the disc, which has an especially dense representation in the visual cortex[30]. This region was included in the present studies and found to show no qualitative differences from other areas. No activity of bipolars has been recorded and therefore all the discharge patterns which are described, while derived from ganglion cells, represent also the discharges in the optic nerve fibres.

Since in each preparation the discharges from numerous units can be observed in quick succession, e.g., 30-40 within an hour, it is possible to collect statistical data on discharge types. It was, however, found more informative to obtain detailed results from a relatively small number of units and frequently these were kept on the electrodes for 5-6 h. Only those cases are presented which, at the present stage, seem more representative or important. The great majority of experiments were done well within photopic levels, with the background illumination between 1 and 50 mc. All the essential features of discharge pattern behaviour, however, were also present under scotopic conditions in the absence of background illumination. It should also be noted that in this study relatively short flashes were used and no 'equilibrium' conditions were attained.

Fluidity of discharge patterns. The most outstanding feature in the present analysis is the flexibility and fluidity of the discharge patterns arising in each receptive field. Stability of discharge type can be obtained in the present preparations in units under certain conditions, especially with a relatively strong background illumination, when the surround is suppressed. A constant 'on' or 'off' response may then be seen even with spot illumination. Such stability, however, disappears when one or more of several parameters, such as the adaptation level, stimulus intensity, and area of illumination, are changed singly or in combination. In the absence of a fixed pattern from the whole receptive field, it does not appear accurate enough to speak of 'on,' 'on-off' or 'off' fibres in the cat's retina. The difference in retinal discharge pattern distribution between frog and cat is worthy of note, particularly since the analyses in frog were made by Hartline[18] with a well-controlled and accurate technique. Although he reported the discharge patterns in receptive fields fixed, he points out many exceptions and reports occasional units in which a change in discharge patterns did occur. He also presents data which may be interpreted to indicate the presence of inhibitory surrounds, such as a decline of discharge frequencies with strong stimuli or with large areas of excitation[17, 19]. The difference between cat and frog may turn out to be largely a quantitative one. A less flexible system of discharge in frogs may perhaps not be surprising. By using a different approach, such as pharmacological techniques[12] and passing current through the eye or varying the wavelength and intensity of the stimulating light, shifts in discharge patterns have already been observed by Granit and his colleagues. They also repeatedly pointed out the lability of certain portions of the discharges, particularly in connection with work concerning the on-off ratios[8, 14, 15]. Donner and Willmer[9] working with dark-adapted cats and stimulation of the whole retina, also observed a great range of variability in ganglion response patterns during stimulation at different intensities. They have shown that visual-purple-dependent receptors can give rise to both 'on' and 'off' discharge components.

Functional organization of receptive fields. There seems to exist a very great variability between individual receptive fields and therefore a detailed classification cannot be made at present. Some features, however, of general organization were found common to all. In all fields there exists a central region giving a discharge pattern which is the opposite from that obtained in the periphery. The centre may be either predominantly 'off', the surround 'on', or vice versa. A transitional zone is in between (see *Figure 6*). The essential character of discharge within the *centres* cannot be changed by altering any of the parameters of illumination, i.e., an 'off' centre cannot be converted into an 'on' centre. It must not be inferred, however, that the centres are quite uniform and receive no contribution which is characteristic of the surround. In view of the fixed nature of the centre discharge, it may be convenient to classify receptive fields into 'on' centre and 'off' centre fields. In line with this the respective elements may be similarly designated as 'on' centre or 'off' centre units. No accurate record of distribution has been made in hundreds of units which were investigated. The 'off' centre units seemed to occur more frequently. Functionally the centre and surround regions are opposed, the one tending to suppress the other. The ganglion cell is subjected to multiple influences from its receptive field and its discharge will express the balance between these opposing and interacting contributions. In view of the relative ease with which the peripheral receptive field contribution can be altered (*see* later), and thereby the balance within the unit changed, the discharge pattern fluidity is readily appreciated.

From a functional point of view, then, the important aspect of the present findings is not that one unit can give under special conditions either 'on' or 'off' responses but that there exists a mixture of contributing receptors, perhaps with their specific pathways (below). In proportion to their activation they can produce all shades of transitions from one response pattern to another. In any event, illumination of the whole receptive field will always produce a push-pull action as the opposing components are thrown into activity.

Specific neural pathways. The nervous organization of all the elements functionally connected to a ganglion cell constitutes an example of the complexity of the central nervous system, well known from the studies of Cajal[5] and lately especially of Polyak[26]. It is natural that a specific organization should be suggested for excitatory and inhibitory pathways for which there is physiological evidence. Experiments seem to show (*Figure 6*) that excitation of a certain number of receptors by restricted illumination causes one type of response only. Presumably a given pathway is utilized by a given group of receptors. The principal reason for a change in response type seems to be that either additional receptors have been brought in or receptors have been eliminated (see below). Suggestions as to the specific neural connections, based on present evidence, are clearly speculative and grossly simplified. One may think of a neural arrangement which parallels the roughly concentric functional pattern, with relatively uniform connection types between receptors in the centre and the ganglion cell and a differing pathway set-up from the surround receptors to the ganglion cell. The in-between region may present the zone where the pathway types are more mixed than anywhere else. A correlation of greater significance between neural pathways and discharge patterns may perhaps be obtained from a study of animals with a fovea, e.g., monkey, where the neuroanatomical connections are simpler and better known. It may be predicted, accordingly, that the foveal paths are associated with specific discharge behaviour.

Receptor density in receptive fields. Any given small area of the retina which has been studied presumably has a dense and fairly uniform receptor population[26]. Histological data also show that adjoining receptors, or even the same receptors, may have connections to different ganglion cells. Further, we know that the same receptor may connect to different ganglion cells in differing ways. This is the neuroanatomical basis for overlapping receptive fields. The present study gives some information about the density distribution of receptors which are *functionally* connected to one ganglion cell. The following type of experiment supports the assumption that the central portion of receptive fields hold a denser population of receptors per unit area than do the peripheral regions: units which have an 'on' centre and 'off' surround, when tested with stimuli 100-1000 times threshold (*Figure 6*), may be excited in their most sensitive central region by a 0·2 mm diameter spot of near-threshold intensity. At this strength 'on' discharges of quite short duration are evoked within a small central area (as *Figure 5*). Such a small spot is well below threshold for the outlying portions of the receptive field. Placing a ground glass in front of the eye, thereby reducing, and in addition scattering, the light beam, will produce 'on-off' or pure 'off' responses. This suggests that receptors dispersed in peripheral regions have been reached and summation has occurred in their pathways leading to the ganglion cell. The experiment also shows that the receptors which contribute the 'off' component do not have a lower threshold than those in the central 'on' region. Threshold differences, however, within the receptor population are not contra-indicated by such results. Presumably because of the density of receptors, the centre is found to be more sensitive when tested with small beams.

The scatter of receptors in the periphery makes obvious the difficulties of receptive field mapping with small light beams, since one has to assume that a sufficient number of receptors must be activated to evoke a ganglion cell discharge. Receptors, functionally connected but located in the periphery, will be missed and an error in underestimating the field size is likely to be made. When mapping is done in units with an 'off' surround, while they show spontaneous activity, the field periphery can be delineated by the area over which a small spot will produce slowing or stoppage of firing. This method is more sensitive than the one described in Section 3 and receptive fields extending over 3-4 mm (12-16 degrees in cat), could occasionally be obtained. The effect of light scatter, however, could not be estimated closely enough to make these findings reliable.

The low density of receptors in the surround also makes readily understandable the observed shrinkage of receptive fields with augmented background illumination. If tested with a *small spot*, the dropping out of receptors by raised thresholds in the field surround will be of more consequence than in the dense central region, since the outer receptor family, being scattered, operates nearer to the margin for firing the ganglion cell. Even with illumination of the *whole field* the peripheral contribution, if it depends more on facilitation and summation, should be more affected if a portion of component pathways is put out of action. The background changes should affect an 'on' or 'off' surround equally, in line with present observations.

Changes in receptive field contribution to ganglion cells. A special nervous organization of receptive fields alone could not account for all the observed discharge pattern changes under diverse conditions of illumination. There is a great body of evidence for a diversity of receptor properties in respect to thresholds, adaptation, wavelengths, etc. For instance, in view of the changes in receptive field size at low or high levels of light adaptation, it is clear that under such changing conditions a largely differing set of receptors will be thrown into activity with a given stimulus. Hence, this alone will bring a different set of active connections with a ganglion cell into being. The differing connections, in turn, are likely to cause changes in discharge patterns. Since steady states cannot be attained, a shift in the active receptor population is likely to go on continuously even in the dark-adapted eye, as indicated by the background activity, unless the latter is entirely due to spontaneous rhythms in the neural elements.

Psychophysical aspects. A transference of information about discharge patterns, as obtained here, to psychophysical data is obviously based on speculation. However, the data must be used with all their imperfections since they provide the components which form the basis of the message content reaching the higher centres. The most potent stimuli, those causing the greatest nervous activity, are relatively sudden changes. These may be either changes of the general illumination level or such changes as occur during movement of images (see also Reference 17). In the latter case the antagonistic arrangement of central and peripheral areas within receptive fields seems important since the smallest shift can cause a great change in the discharge pattern. This should be advantageous in the perception of contrast and in acuity. In view of this the importance of small eye displacements, as would occur in any scanning movement, is clear. It should be noted that zonal gradients within fields, between centre and surround, will change with different levels of adaptation as the receptive field shrinks or expands. It may also be tempting to consider in this connection well-known suppression phenomena like lateral inhibition which has been studied by many investigators. Particularly the interaction of two light patches with facilitation and inhibition as observed in humans over small distances on the retina may be considered in view of the cat's receptive field[10, 11]. It is clear, however, that even the smallest light beams used in the present experiments do excite a great number of ganglion cells through their overlapping receptive fields. It is not known how the latter are functionally related to each other. For instance, it would be of interest to know whether the same receptor can be connected to one ganglion cell through an inhibitory pathway.

Regarding the information content carried by a single ganglion cell or its connected nerve fibre, the following two phenomena may be briefly considered: (i) if a ganglion cell can discharge at one time during illumination and then be converted into one which signals only when a light stimulus is withdrawn, one may assume that it does not carry merely the information which is suggested by a 'simple' interpretation of the discharge pattern. The higher centres may receive identical impulse patterns in both cases. (ii) A unit giving 'on-off' discharges in a given situation, according to a 'simple' interpretation, sends information first about an increased and then about a decreased level of luminosity. The identical discharge pattern may, however, be evoked by turning a light on, and then instead of turning it off, one may further increase its intensity. Such fibres then merely signal change, such as brightness or darkness.

In view of the massive continued nervous activity in eyes 'at rest' or during illumination it is difficult to think of information content in terms of single unit contributions. One may rather have to consider that groups of fibres modulate activity levels and patterns by superposition or subtraction. The latter—for instance, transient cessation or diminution—is likely to be as meaningful as the opposite in terms of message content. Further, similar discharge patterns at different background illuminations may convey a different meaning, since they are superimposed on a different background activity. These examples merely illustrate some of the difficulties inherent in this type of analysis. They indicate that on the basis of the single unit discharge a 1:1 agreement between discharge patterns and information should not always be sought. At the same time it should be recalled that there is agreement between psychophysical measurements such as the visibility curves and the analogous curves, obtained in different mammals[13] or the Limulus[20] from nervous discharges.

In this study the influence of transient light changes on discharge patterns has been emphasized; in view of the importance of background activity the effect of steady levels of illumination on discharge behaviour must be analysed in great detail.

SUMMARY

Discharge patterns from the unopened cat's eye have been studied by recording from single cells in the retina. Small electrodes, inserted behind the limbus, traversed the vitreous and made light contact with different regions of the retina. The normal optics of the eye were used for stimulation by two independently controlled light beams. Circular stimuli of various dimensions, duration and intensities were applied to different areas of the retina. A third light source provided the background illumination, determining the adaptation level, and also served for simultaneous direct observation of the fundus.

1. The discharges arising in nerve fibres and ganglion cells can be readily distinguished through differences in their time course and the location of their respective receptive fields. The present study was done on ganglion cell activity.

2. Ganglion cells can be blocked reversibly by pressure and the potentials can be split into 'prepotentials' and 'spikes'.

3. Under dim background illumination and during dark adaptation the cat's retina is dominated by generalized spontaneous activity. The latter is reduced by illumination and anaesthetics such as Dial or Nembutal. Certain discharges do not seem to be

influenced appreciably by illumination. These observations are in general agreement with Granit's findings.

4. The configuration of receptive fields—those areas of the retina which must be illuminated to cause a discharge in a ganglion cell—were studied by exploration with small spots of light. The fields are usually concentric, covering an area of 1-2 mm, or possibly more, in diameter. The boundaries and extent of receptive fields cannot be delineated accurately. They shrink under high background illumination and expand during dark adaptation.

5. The discharge pattern from individual ganglion cells is not fixed. 'On', 'off' or 'on-off' discharges can be obtained from one ganglion cell if specific zones within its receptive field are stimulated by small spots of light. The discharge pattern from a ganglion cell depends, amongst others, on the following factors: background illumination and the state of adaption, intensity and duration of stimulation, extent and location of area which is stimulated within a receptive field. Each of these parameters can alter the discharge pattern by itself or in conjunction with the others.

6. The general functional organization of each receptive field is the following: there exists a central area of low threshold as tested by a small spot of light. The discharge pattern of the central region is the opposite of that found in the periphery or surround. The centre may give predominantly 'off', the surround 'on' discharges, or the reverse. An intermediary region gives 'on-off' discharges. The units which carry discharges from 'centre on' or 'centre off' receptive fields may accordingly be classified as 'on' centre or 'off' centre units. A conversion of one type into another by changing conditions of illumination has not been possible.

Experiments indicate that receptors in the periphery of receptive fields are less dense per unit area than in the central regions.

7. Interaction of different regions within single receptive fields was studied by simultaneous excitation by two small beams of light. Depending on a number of factors, 'off' areas may suppress the discharge from 'on' regions, or vice versa. All degrees of mutual modification can be obtained. It is assumed that specific areas give rise to predominantly inhibitory or excitatory pathways to a given ganglion cell.

8. The discharge pattern of a ganglion cell, set up by illumination of an entire receptive field, depends on the summed effects of interacting pathways converging on the cell. The ratio of functionally opposing centre and surround regions varies greatly

in different receptive fields. Under diverse conditions of illumination the balance of active inhibitory and excitatory contributions changes within the same receptive field. This seems to be responsible for the varied discharge patterns.

9. The character of 'on' components in discharge patterns was also studied. The maintained 'on' response discharges for the duration of illumination while the transient 'on' adapts quickly. Transitions between these two 'on' types were found in the same receptive fields. It is suggested that the transient 'on' discharges are the result of various amounts of addition by the 'off' surround to the 'on' centre.

10. The latent periods of 'on' or 'off' responses were found to be shorter than hitherto observed, presumably because of more exclusive stimulation of their specific receptive areas. High discharge frequencies of 200-800/s were found to be within the normal range in the cat's eye.

11. Some 'anomalous' observations, such as lengthened latent periods with increased stimulus intensities or higher frequency discharges with weaker stimuli, are interpreted in the light of receptive field organization.

ACKNOWLEDGMENT

I am grateful to Dr. S. A. Talbot for his help, particularly in the design of the optical and electronic instruments, which made this study possible. My thanks are also due to Mr. Albert Goebel for constructing the optical apparatus.

REFERENCES

[1] Adrian, E. D. (1937). 'Synchronized reactions in the optic ganglion of Dytiscus.' *J. Physiol.*, *91:* 66-89

[2] Adrian, E. D. and Matthews, R. (1927). 'The action of light on the eye. Part I. The discharge of impulses in the optic nerve and its relation to the electric changes in the retina.' *J. Physiol.*, *63:* 378-414

[3] Adrian, E. D. and Matthews, R. (1928). 'The action of light on the eye. Part III. The interaction of retinal neurones.' *J. Physiol., Lond.* . *65*, 273-298.

[4] Bernhard, C. G. (1940) 'Contributions to the neurophysiology of the optic pathway.' *Acta physiol. scand.*, *1:* Suppl. 1.

[5] Cajal, S. Ramón Y. (1933). 'La rétine des vertébrés.' *Trab. Lab. Invest. biol. Univ. Madr.*, Suppl. 28.

[6] Chievitz, J. H. (1891). 'Über das Vorkommen der Area centralis retinae in den vier höheren Wirbelthierklassen.' *Anat. entwgesch. Monogr.*, pp. 311-334

[7] Crescitelli, F. and Jahn, T. L. (1939). 'The effect of temperature on the electrical response of the grasshopper eye.' *J. cell. comp. Physiol.*, *14*, 13-27

[8] Donner, K. O. and Granit, R. (1949). 'The effect of illumination upon the sensitivity of isolated retinal elements to polarization.' *Acta physiol. scand.*, *18:* 113-120.

[9] Donner, K. O. and Willmer, E. N. (1950). 'An analysis of the response from single visual-purple-dependent elements in the retina of the cat.' *J. Physiol., Lond.*, *111*, 160-173

[10] Fry, G. A. (1934). 'Depression of the activity aroused by a flash of light by applying a second flash immediately afterwards to adjacent areas of the retina.' *Am. J. Physiol.*, *108*, 701-707.

[11] Granit, R. (1930). 'Comparative studies on the peripheral and central retina. I. On interaction between distant areas in the human eye.' *Am. J. Physiol.*, *94*, 41-50.

[12] Granit, R. (1945). 'Some properties of post-excitatory inhibition studied in the optic nerve with micro-electrodes.' *K. svenska Vetensk-Akad. Arkiv. Zool.*, *36*, 1-8.

[13] Granit, R. (1947). *Sensory mechanisms of the retina.* London; Oxford Univ. Press, 412 pp.

[14] Granit, R. (1948). 'Neural organization of the retinal elements, as revealed by polarization.' *J. Neurophysiol.*, *11*, 239-253.

[15] Granit, R. (1950). 'The organization of the vertebrate retinal elements. *Ergebn. Physiol.*, *46*, 31-70.

[16] Granit, R. and Therman, P. O. (1935). 'Excitation and inhibition in the retina and in the optic nerve.' *J. Physiol., Lond.* 83, 359-381.

[17] Hartline, H. K. (1938). 'The response of single optic nerve fibers of the vertebrate eye to illumination of the retina.' *Am. J. Physiol.*, *121*, 400-415.

[18] Hartline, H. K. (1940). 'The receptive field of the optic nerve fibers.' *Am. J. Physiol.*, *130*, 690-699.

[19] Hartline, H. K. (1940). 'The effects of spatial summation in the retina on the excitation of the fibers of the optic nerve.' *Am. J. Physiol.*, *130*, 700-711.

[20] Hartline, H. K. (1940). 'The nerve messages in the fibers of the visual pathway.' *J. opt. Soc. Am.*, *30*, 239-247.

[20a] Hartline, H. K. (1941). 'The neural mechanisms of vision.' *Harvey Lect.*, *37*, 39-68.

[21] Hartline, H. K. and Graham, C. H. (1932). 'Nerve impulses from single receptors in the eye.' *J. cell. comp. Physiol.*, *1*, 277-295.

[22] Hartline, H. K., Wagner, H. G., and McNichol, E. C. (1952). 'The peripheral origin of nervous activity in the visual system.' *Cold Spr. Harb. Symp. quant. Biol.*, 17.

[23] Hodgkin, A. L. (1938). 'The subthreshold potentials in a crustacean nerve fiber.' *Proc. Roy. Soc.*, *B126:* 67-121.

[24] Hunt, C. C. and Kuffler, S. W. (1951). 'Stretch receptor discharges during muscle contraction.' *J. Physiol., Lond.* 113, 298-315.

[25] Katz, B. (1937). 'Experimental evidence for a non-conducted response of nerve to sub-threshold stimulation.' *Proc. Roy. Soc.*, *B124*, 244-276.

[26] Polyak, S. L. (1941). *The retina*. Chicago; Univ. of Chicago Press, 607 p.

[27] Renshaw, B. (1942). 'Effects of presynaptic volleys on spread of impulses over the soma of the motoneuron.' *J. Neurophysiol.*, *5*, 235-243.

[28] Rushton, W. A. H. (1949). 'The structure responsible for action potential spikes in the cat's retina.' *Nature, Lond. 164*, 743-744.

[29] Svaetichin, G. (1951). 'Analysis of action potentials from single spinal ganglion cells, *Acta physiol. scand.*, *24*, Suppl. 86.

[30] Talbot, S. A. (personal communication).

[31] Talbot, S. A. and Kuffler, S. W. (1952). 'A multibeam opthalmoscope for the study of retinal physiology.' *J. opt. Soc. Am.*, Dec. (in press).

14

CODING AND INFORMATION THEORY

Of all the disciplines that have contributed to our understanding of biological systems Information Theory is the most abused. This short paper gives a useful introduction to the ideas of communication theory and to the concept of 'information content'. It is pedagogical in intention: there is no original content. None the less, an understanding of Shannon's mathematical theory of communication is essential for any insight into the biological control systems which we are attempting to model. Words like 'information', 'coding', 'noise' and so forth are consistently misused in the biological literature: indeed they sometimes seem to have a special magic, very seductive to the biologist who has no real understanding of the problems confronting him.

Further Reading

Pierce, J. R. (1962). *Symbols, Signals and Noise*, Hutchinson
Winograd, S. and Cowan, J. D. (1963). *Reliable Computation in the Presence of Noise*, M.I.T. Press

Reproduced from *Biophysical Science: a Study Program* (Oncley Ed.) (1959) pp. 221-226 by courtesy of Wiley.

CODING AND INFORMATION THEORY

Peter Elias

INTRODUCTION

'Information theory' is used in at least three senses. In the narrowest of these senses, it denotes a class of problems concerning the generation, storage, transmission, and processing of information, in which a particular measure of information is used. This area is also called 'coding theory' and, especially in Britain, 'the mathematical theory of communication', which is the title of Shannon's original paper[1] from which the field is derived. This is the usage in the titles of the books by Khinchin[2] and Feinstein[3], which are rather abstract mathematical presentations.

In a broader sense, information theory has been taken to include any analysis of communications problems, including statistical problems of the detection of signals in the presence of noise, that make no use of an information measure. Woodward[4] has shown the relationship of information measure to some of these problems. A book by Weiner[5], an article by Rice[6,7] (reprinted in a book edited by Wax[8]), and a recent textbook by Davenport and Root[9] discuss the use of statistical techniques in problems concerning analysis of signals and noise, and books by Blanc-LaPierre and Fortet[10], Loeve[11], and Doob[12] provide the (abstract) mathematical background. It is in this broader sense that the word is used, for example, in the title of the Professional Group on Information Theory of the Institute of Radio Engineers, whose *Transactions* contains articles both on coding and on signal-noise problems.

In a still broader sense, information theory is used as a synonym for the term 'cybernetics' introduced by Wiener[13] to denote, in addition to the areas listed in the foregoing, the theory of servomechanisms, the theory of automata, and the application of these and related disciplines to the study of communication, control, and other kinds of behaviour in organisms and machines. This is the usage in the titles of three meetings held in London (two of which have published proceedings[14,15]) and two held at the Massachusetts Institute of Technology[16,17] *Information and Control* publishes articles in this broad area.

On¹y the area covered by the narrowest definition is discussed here—not because it is necessarily the most important for biologica¹ applications, but because of the limitations of space.

INFORMATION MEASURE SOURCES

The first problem is to assign a measure to information. *Figure 1* shows two representations of a message and the code book which connects them. At first one might suppose that a long message has more information than a short one, but, as the figure shows, the message length—or any other characteristic of the representation itself—can be changed drastically by coding at the transmitter. If the receiver decodes correctly, the message has been transmitted successfully by using a very brief form. In a communications system, any message has a variety of different representations in different places. One may want to say that it contains the same amount of information. An amount of information, then, cannot depend upon the form of representation. It seems reasonable, however, to make it depend upon how many messages there are. If the number of messages is large, longer code words will have to be used in order to distinguish between them. One starts, therefore, with the hypothesis that the information in a message is some function of the number of messages in the set. One would like to say that two successive, independent selections from the same set have twice the information value of a single selection, *if the two messages are equally probable.* This demands a function $f(m)$ of the number m of messages in the set, for which $f(m^2) = 2f(m)$, and the logarithm is the only respectable function with this property. For a selection between two messages, this gives $[\log 2]$ units of information. This can be rewritten as $[-\log (\frac{1}{2})]$ or, in general, as the negative of the logarithm of the message probability. This, in fact, is what one chooses to generalize as the information associated with a message. The choice of a logarithmic base determines the unit of information, and the base 2 is chosen, the unit being the 'bit'. Thus, a selection between two equiprobable alternatives requires one bit of information.

This definition is plausible, but it does not justify the information measure. The adopted measure is justified, in a practical sense, on the average, because a source that generates infor-

mation more rapidly than another, also requires more communications facilities—more bandwidth, time, signal-to-noise ratio—
to transmit its output successfully.

The information source, *Figure 1*, is characterized by the
distribution of information, *Figure 2*. Since this source generates
one bit of information per selection regardless of which message
is selected, its information distribution is degenerate: all of the
probability is piled up at one bit. Each symbol has a probability

CODED MESSAGES
ABBA...

THE QUICK BROWN FOX JUMPED OVER THE LAZY DOG NOW
IS THE TIME FOR ALL GOOD MEN TO COME TO THE AID OF
THE PARTY NOW IS THE TIME FOR ALL GOOD MEN TO
COME TO THE AID OF THE PARTY THE QUICK BROWN FOX
JUMPED OVER THE LAZY DOG...

CODE BOOK

$A \leftrightarrow$ THE QUICK BROWN FOX JUMPED OVER THE LAZY DOG
$B \leftrightarrow$ NOW IS THE TIME FOR ALL GOOD MEN TO COME TO THE
    AID OF THE PARTY
$I(A) = I$(THE QUICK BROWN FOX JUMPED OVER THE LAZY
    DOG) $=$ LOG $2 = -$ LOG$(1/2) = -$ LOG $Pr[A] = 1$ BIT/SYMBOL

Figure 1. Two representations of a message and the code book which connects
them

given by the value of the exponential at the information value of
the symbol, as the plot is just probability *vs* minus log probability. *Figure 3* shows the information distribution for another
two-symbol source whose symbols have the unequal probabilities
¾ and ¼. This source generates only 0·42 bits when it selects
an *A* and 2·0 bits when it selects a *B*. Its average rate is 0·81
per symbol, which is smaller than in *Figure 2* with equiprobable
symbols. This is a general characteristic of sources. Any constraint on the number of symbols or sequences that a source
may generate and any shift away from equal probabilities will
reduce the average source rate. A source constrained to generate
sequences of letters which spell out sentences in English has a
lower average rate than a source which selects successive
letters of the alphabet with statistical independence.

There are, of course, many questions which arise in connection
with coding and with the description of sources having sequential
constraints. What has been covered here is a mathematical theory,
and problems arise in its application. One of these is illustrated

in *Figure 4*. The problem consists in identifying the alphabet which is relevant in a given situation. Suppose that, in observing the output of a neuron, the eight wave forms shown in the figure are seen with equal frequency. This would give three bits of

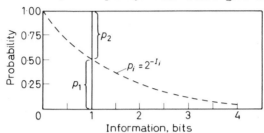

Figure 2. Distribution of information for a two-symbol source whose symbols have equal probabilities

$$p_1 = p_2 = \tfrac{1}{2} \quad I_1 = I_2 = -\log\tfrac{1}{2} = 1 \text{ bit}$$

$$\bar{I} = H = \tfrac{1}{2} + \tfrac{1}{2} = 1 \text{ bit}$$

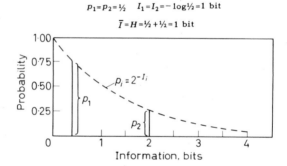

Figure 3. Distribution of information for a two-symbol source whose symbols have the unequal probabilities ¾ and ¼

$$p_1 = \tfrac{3}{4} \quad I_1 = -\log\tfrac{3}{4} = 0 \cdot 42 \text{ bit}$$

$$p_2 = \tfrac{1}{4} \quad I_2 = \log\tfrac{1}{4} = 2 \cdot 00 \text{ bits}$$

$$\bar{I} = \sum_{i=1}^{2} p_i I_i = H = \sum_{i=1}^{2} -p_i \log p_i = 0 \cdot 31 + 0 \cdot 50 = 0 \cdot 81 \text{ bits}$$

information per wave form. However, these signals may not all be distinguishable by the system being observed. The system may act only as a pulse counter over this time interval, and may recognize only four different signals—no pulse, one pulse, two pulses, and three pulses. Then the average rate is only 1·82 bits per signal. There is no way of telling from the outside which of these alphabets is actually in use—if, in fact, either is. It is necessary to determine that the system does have different responses to two signals before they can be defined as being distinct letters of an alphabet. The problem of recognizing the relevant alphabet always is present and shows up in many ways, including the selection of scales of resolution to be used in amplitude and time to distinguish different signals. In *Figure 4*, the average rates for the two alphabets are not too far apart, but in a train of 100 pulses, if all observed wave forms are equiprobable, there is a factor of about 20 between the rates obtained for the two corresponding alphabets.

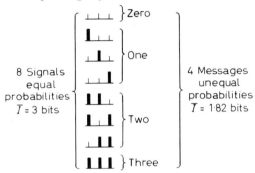

Figure 4. Dependence of source rate on alphabet

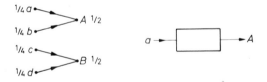

Figure 5. The mutual information between a transmitted symbol and a received symbol

$$Pr(a) = \tfrac{1}{4} \qquad Pr(a \mid A) = \tfrac{1}{2}$$

$$I(a) = 2 \text{ bits} \qquad I(a \mid A) = 1 \text{ bit}$$

$$\Delta I = \log Pr(a) - \left[-\log Pr(a \mid A) \right] = 1 \text{ bit}$$

$$= \log \frac{Pr(a \mid A)}{Pr(a)} = \log \frac{Pr(a,A)}{Pr(a)Pr(A)} = I(a; A)$$

NOISY CHANNELS

There are then, sources to select symbols, and channels are needed to transmit them. *Figure 5* shows one—a noisy channel. In communications, a noisy channel usually is a medium which

257

separates transmitter and receiver. In information storage, a noisy channel may model the action of the environment which, through thermal agitation or other forces, may cause changes in the stored information. In *Figure 5*, the channel eliminates some of the distinctions present in its input. The source selects from among four equiprobable symbols a, b, c, and d. The receiver on reception of A or B can eliminate two input possibilities, but it cannot choose between the other two.

To analyse informationally what happens in the channel, consider a particular transmission event: the selection of a at the transmitter and the reception of A at the receiver. It is assumed that the receiver knows the input-letter probabilities and the channel. Thus, before reception of A, the estimate by the receiver of the probability that a will be transmitted is ¼. After reception of A, the receiver knows that only either a or b could have been transmitted, and both are equally likely. Thus, *a posteriori*, the probability of a (on the evidence A) is ½.

The receiver *a priori* needs log 4 = 2 bits of information to select a as the transmitted letter. A posteriori, it still needs log 2 = 1 bit to select a after receiving A. The channel is credited with the 1-bit difference, which is defined as the amount of information which the receipt of A gives about the transmission of a. The quantity expressed in *Figure 5* as $I(a; A)$ is called the *mutual information* of A about a, or the *transmitted information*.

In general, for each possible pair of transmitted and received symbols, x and y , there is a probability of occurrence $Pr(x_i, y_j)$ [which may be zero as in *Figure 5* for $Pr(a, B)$] and an information value $I(x_i, y_j) = \log Pr(x_i, y_j)/Pr(x_i)Pr(y_j)$, which measures the change in the logarithm of the probability of x_i owing to knowledge of y_j. This quantity is positive if y_j makes x_i more probable than it was, and is negative if y_j makes x_i less probable than it was. A plot can be made of the *mutual-information distribution* of a channel-source combination. For the channel of *Figure 5*, this plot is exactly like the plot of *Figure 2*: one bit of information is always transmitted regardless of which input-output pair happens.

Two more-complicated channels and their mutual-information distributions are shown in *Figure 6*. In the binary erasure channel (BEC), one of two symbols is selected at the transmitter. The input symbol may be received correctly, with probability q taken here as ¾, or it may be erased with probability $p = ¼$: The receiver then receives an X. Evaluating mutual information gives the distribution shown. The channel sends one bit per symbol three-quarters of the time, when no erasure occurs; one-quarter of the time, when the transmitted symbol is erased, no information is transmitted. The average rate is just ¾ bit per symbol, if the input symbols are equiprobable.

In the binary symmetric channel (BSC), there are only two input and two output symbols, and true errors occur. Here less than one bit is transmitted when there is no error, and a negative amount of mutual information is transmitted if there is an error.

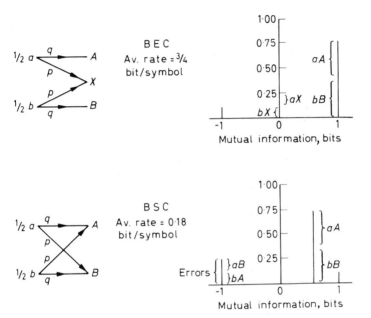

Figure 6. The binary erasure channel, the binary symmetric channel, and their mutual information distribution

One takes error probability $p = ¼$. The average rate of transmission here is only 0·18 bits per symbol with equiprobable inputs.

Returning to *Figure 5*, one interprets the average rate of transmission of 1 bit per symbol as the average rate at which the receiver's ignorance is reduced—from 2 bits to 1 bit for each letter transmitted. However, the receiver's ignorance is not reduced to zero. There is an apparent difference between this channel and another with the same rate which transmits two equiprobable symbols without error. In the latter case, the ignorance of the receiver is reduced from 1 bit to zero bits for each transmission. One would like to be able to say that the average rate over a channel is the significant parameter. This requires that, for the channel of *Figure 5*, one finds some method for which the rate of putting information in is reduced in such a way as to reduce both the initial and final ignorance of the receiver, keeping their difference fixed. In this case, it is easy to do. The transmitter agrees to send only two symbols, a and c, with equal probabilities. The receiver then can decode unambiguously, receiving one bit of mutual information per symbol with no residual ignorance.

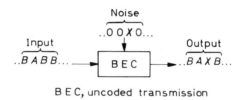

BEC, uncoded transmission

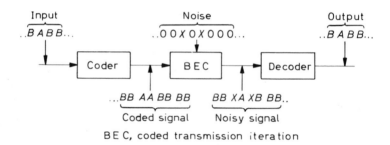

BEC, coded transmission iteration

Figure 7. Transmission over the binary erasure channel uncoded and coded by iteration

CODING FOR NOISY CHANNELS

The foregoing result is, in fact, general for a broad class of noisy channels, although the way in which the input information rate is reduced is usually more complicated. It is not possible in the BEC or BSC to find a set of input symbols which will leave the receiver with no ambiguity. All input symbols are used, therefore, but they are used in sequences, and only a fraction of the possible sequences occur.

Figure 7 shows the BEC, first used in ordinary, uncoded transmission and then used with a simple coding and decoding scheme. In the upper picture, one bit per symbol is put into the channel, ¼ of the output symbols are erased, and an average of ¾ bit per symbol comes out. Below, a coder is added, which duplicates each input symbol. For a fixed channel, input symbols can be accepted now only half as often, since two symbols go into the channel for each input symbol to the coder. The input rate is then ½ bit per channel symbol. Again ¼ of the output symbols are erased, but both of the copies of an input symbol are erased $\frac{1}{16}$ of the time only, so reliability has improved.

Of course, each digit could be sent three or more times instead of twice. This would reduce further both the rate of transmission and the probability of total erasure, or of ambiguity after de-

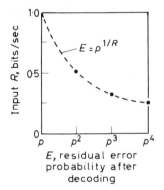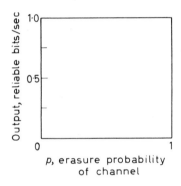

Figure 8. Rate and error probability for iteration coding of the BEC

coding. The relationship is shown in *Figure 8*. On the left plot, it can be seen that sufficient reduction of rate reduces residual probability of total erasure as low as is desired, but, to get arbitrarily low probability, an arbitrarily low rate of transmission is necessary. If one looks at the channel in a different way, demanding transmission with *arbitrarily low* total-erasure probability, and asking how average rate varies as the channel-erasure probability p is varied, one sees the plot on the right in *Figure 8*. If $p = 0$, one can send one reliable bit per symbol, but if $p \neq 0$, none can be sent. Time must be spent on repeating the first input symbol to make its reliability arbitrarily good, and one never can get around to sending anything else.

Certainly, one way of avoiding misunderstanding is never to say anything new, but it would be discouraging if this were the only way. An alternative is shown in *Figure 9*.

Here one starts with uncoded transmission again. Now, however, instead of duplicating each input symbol, a general purpose replacement, called a check digit or a parity check, is inserted so that a replacement will be available in case it is erased. The digit enclosed in the dotted box is selected to make the

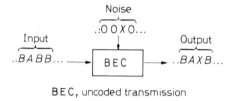

BEC, uncoded transmission

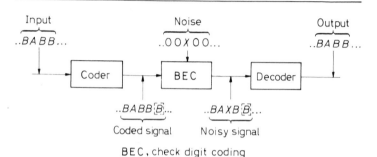

BEC, check digit coding

Figure 9. The transmission erasure channel, the binary symmetric channel, and their mutual information distribution with check digits

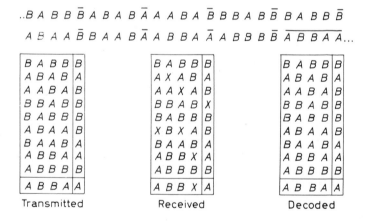

Figure 10. Correction of erasures by iteration of parity-check digits

total number of *B*'s in the sequence of 5 symbols an *even* number. At the receiver, if only one symbol has been erased, the knowledge that an even number of B's should be present makes the decoding unique. The rate of transmission has been reduced to $\frac{4}{5}$ bit per symbol and increased reliability is obtained. There is still a danger, however, that two or more erasures may occur in the same block of five digits, in which case decoding still would be ambiguous.

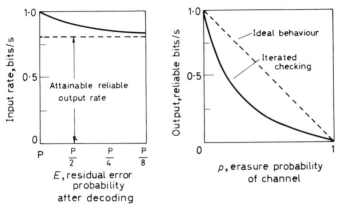

Figure 11. Rate and error probability for iterated parity checking

The situation can be improved by adding further check digits as shown in _Figure 10_. Here the sequence of symbols is shown above. Added check digits have bars over them. The check digits are computed as shown below: each symbol to the right of a row is selected to make the total number of B's in the row even; each symbol at the foot of a column is selected to make the total numbers of B's in the column even. First, those rows with single erasures are decoded, and then the columns having only single erasures remaining after row correction. In the case illustrated, this corrects all erasures. Further high-order check digits can be added indefinitely to give a total-erasure probability which is arbitrarily small without reducing the transmission rate to zero. _Figure 11_ shows the kind of relationship between rate and residual-erasure probability which results. Here an erasure probability p of 1/20 and an initial check group of ten, rather than of five, symbols have been used. Rate still goes down as reliability increases, but it now has a positive asymptote at 0·80 bits per symbol, at which rate it is possible to get arbitrary reliability. The plot of reliable rate vs the channel-erasure probability p is now continuous, as shown by the solid curve. This kind of iterated check-symbol coding can be used for the BSC[18] as well as for the BEC[19].

Although a very special case of noisy-channel coding has been shown, it illustrates a general result. Given any noisy channel and source, with an average rate of transmission, it is possible to code the input in long blocks and to reduce both the input rate and the receiver's residual uncertainty, while keeping the average transmission rate fixed. The ideal behaviour is illustrated in the dotted line on *Figure 11* for the BEC; to approach it, coding in large blocks is required.

One other point should be noted. The code we constructed was engineered carefully, and its reliability may appear to be atypical. In fact, by random selection of long sequences of symbols, it is possible to get the same results—the 'ideal behaviour' of *Figure 11* is obtained by just such random coding. It is necessary only to *not* select too many possible input sequences. Thus, a code suitable for reliable transmission over a noisy channel might occur quite accidentally. However, no accident seems likely to account for the necessary decoding equipment and its organization.

CONCLUSION

In conclusion, it might be appropriate to point out what applications have been made to biological problems. The most successful ones, I think, have been to experiments in human communication where the channel capacity of a human being for handling information has proved a useful concept in a number of experimental situations [see e.g., Rosenblith (p. 485) and reference 20]. There also have been applications at the neurophysiological level, but in terms of statistical signal analysis rather than of information theory *per se;* some of these are referred to in the second paper by Rosenblith (p. 532). Computations of neuron channel capacity have been made, but they are of dubious value in view of the alphabet problem illustrated in *Figure 4*. It seems likely that real applications are forthcoming at this level.

There have been applications to chemical specificity, etc., in biological systems (see, e.g., references 21 and 22). My feeling is that these use information measure either as a language for the discussion of purely combinatorial problems or as a useful statistic, but they do not use it in any coding sense which would imply that the informational treatment was at all necessary or unique. I think that the only other immediate application might arise in connection with the genetic-coding problem. Here, the most urgent need obviously is more data about nucleotide-amino group correspondences and the statistics of series of each. Although informational ideas may be useful here, it seems unlikely that they are essential. That is, it seems unlikely that high orders of redundancy and error-correction are being used. Some data on how local the coding is would be useful in reaching a decision on this point.

REFERENCES

[1] Shannon, C. E., (1949). *Bell Syst. Tech. J.* 27, 379, 623 (1948); reprinted in C. E. Shannon and W. Weaver's *The Mathematical Theory of Communication*. Urbana, Illinois; The University of Illinois Press.

[2] Khinchin, A. I. (1957). *Mathematical Foundations of Information Theory*. New York; Dover Publications.

[3] Feinstein, A. (1958). *Foundations of Information Theory*. New York; McGraw-Hill.

[4] Woodward, P. M. (1953). *Probability and Information Theory*. New York; McGraw-Hill.

[5] Wiener, N. (1949). *The Extrapolation, Interpolation and Smoothing of Time Series*. Technology Press, Cambridge, Massachusetts and John Wiley, New York.

[6] Rice, S. O. (1944). *Bell Syst. Tech. J.* 23, 282.

[7] Rice, S. O. (1945). *Bell Syst. Tech. J.* 24, 46.

[8] Wax, N. (1954). editor, *Noise and Stochastic Processes*. New York; Dover Publications.

[9] Davenport, W. B. Jr., and Root, W. J. (1958). *An Introduction to the Theory of Random Signals and Noise.* New York; McGraw-Hill.

[10] Blanc-LaPierre, A. and Fortet, R. (1953). *Theorie des fonctions aléatoires*. Paris; Masson et Cie.

[11] Loeve, M. (1955). *Probability Theory*. New York; D. Van Nostrand.

[12] Doob, J. L., (1953). *Stochastic Processes*. New York; John Wiley.

[13] Wiener, N., (1948). *Cybernetics*. Cambridge, Massachusetts; Technology Press and New York; John Wiley.

[14] Jackson, W., (1953). editor, *Proceedings of a Symposium on Application of Information Theory, London, 1952*. London; Butterworths.

[15] Cherry, C., (1956). editor, *Information Theory, Proceedings of a Symposium*. London; Butterworths.

[16] IRE Trans. Professional Group on Information Theory *PGIT-4* (1954).

[17] IRE Trans. on Information Theory *IT-2* No. 3 (1956).

[18] Elias, P., (1954). IRE Trans. Professional Group on Information Theory *PGIT-4* 29.

[19] Elias, P., (1958). in *Handbook of Automation, Computation and Control*, E. Grabbe, editor. New York; John Wiley. Vol. I, p. 16-01.

[20] Quastler, H., (1955). editor, *Information Theory and Psychology*. Glencoe, Illinois; Free Press.

[21] Quastler, H., (1953) editor, *Essays on the Use of Information Theory and Biology*. Urbana , Illinois; The University of Illinois Press.

[22] Yockey, H. P., Platzman, R. L. and Quastler, H., (1958) editors, *Symposium on Information Theory in Biology*. New York; Pergamon Press.

15

TWO EARS–BUT ONE WORLD

In the perception of the events of the external world we form internal images of the information presented to us within the various sensory modalities. In this paper Cherry develops a model to describe the formation of one kind of sensory image, the 'fusion' of sounds presented to both ears. The fusion process is susceptible to detailed analysis because the signals arriving at either of a subjects' ears can be varied independently and with great accuracy. The model expressed involves auto-correlation of the signal at each ear as well as close-correlation of the two products. Such a system accounts quite well for several classes of auditory phenomena and can predict the effects of change in the pitch and delay between two sounds presented to either ear. The probability of fusion and the probability that a sound source will be heard to the right or to the left of the subject can be calculated.

This type of modelling is useful because not only does it allow us to approach, with increasing precision an adequate description of a system, but it also gives us insight into the sorts of operation necessary for the normal functioning of the system, and therefore may provide some clues as to the function of physical structures known to be present and suggest meaning for the observed neurophysiology.

Further Reading
Littler, T. S. (1966). *The Physics of the Ear*, Pergamon
Sayers, Bruce McA. and Cherry, E. C. (1957). 'Mechanism of binaural fusion in the hearing of speech' *J. acoust. Soc. Am.* **29**, 9, 973-987

TWO EARS—BUT ONE WORLD

Colin Cherry

THE FORMATION OF BINAURAL *GESTALTEN*

I believe it was Epictetus the Stoic who was reported as saying, 'God gave man two ears, but only one mouth, that he might hear twice as much as he speaks.'

This wishful thought may be empty, so far as human wisdom is concerned, but it is curiously near truth on the plane of psychophysics. For the possession of two ears gives us greatly enhanced powers of aural discrimination: we can the better separate a single voice in a buzz of conversation and attend to it; or we can single out a voice from the sounds of traffic, or of the wind, and all the myriad of disturbing noises. The brain takes maximum advantage of the slight differences between the signals that reach the two ears—differences in timing, in intensity, and in microstructure—and by processes of inductive inference breaks down the complex of sounds into separate coherent images, *Gestalten*, which becomes projected to form the subjective 'spatial world' of sound.

The basic fact about any one of these images, say that of a voice, is that it is single; with two ears we hear only one world. By what logical processes does the brain examine and analyse the sense data reaching the two ears, so as to achieve this fusion? Understanding of the fusion process seems to be fundamental to an understanding of directional hearing. Understanding of the logic of the processes whereby voices in a crowd can be separated is a parallel problem; for we *can* separate voices, and other *Gestalten*, with one ear alone but with reduced effectiveness.

In this chapter we shall be concerned only with this basic phenomenon of fusion, discussing it as a psychological problem and its description as a logical (mathematical) problem. Physiology will not be mentioned, for we shall not look inside the head at all. On the contrary, we shall try to set up a model to describe what 'must' be there. Before you physiologists charge me with arrogant presumption in saying this, I should ask you to remember two things. First, we are not saying what physiological *mechanism* must be there, but only by what logical principles it can be described as carrying out its functions (for example, whether it adds or multiplies or integrates); second, if you disagree with any aspects of our mathematical model, it is up to you to show that the logical operations are *physiologically* impossible. If you succeed, we on our side will retract and re-examine our logic (or, more likely, our premises).

The model of binaural fusion, to which I refer here, has been described elsewhere in complete detail (Sayers and Cherry[1]). It is largely the result of the extensive experimental and computational work of Sayers while he was a student in my group at Imperial College, London.

TO FUSE, OR NOT TO FUSE?

In order to study fusion, as opposed to directional hearing (Cherry and Sayers[2]), we eliminate the effects of head turning by fitting the subject with headphones. If the two earphones are driven with identical signals of any kind, the subject hears a single fused image located in the centre of his head. If the inter-aural time difference T_e, or the relative intensity A_L/A_R, of the left- and right-hand signals is varied, the image appears to move across the head laterally in a line between the two ears. The image does not appear to pass outside the head and has no angular direction, as in real life; it has only lateral position, left to right.

We control the lateralization by varying the interaural time interval T_e only, driving each earphone from a separate repro-ducing head on a special magnetic-tape recorder, running at high speed. The interaural time interval T_e can be set to any value to an accuracy of less than 20 μs.

The interval T_e is set to a succession of random values (using a table of random numbers), and the subject is required to guess whether the fused image appears to lie to the right or left—a dichotomous judgment, R or L. The forms of such lateralization 'judgment curves' fall broadly into two classes: (1) If the source is predominantly random, or quasi-random, the form is like that shown in *Figure 1a*. When T_e is large (say 1 to 5 ms), the subject answers 100 per cent correctly; when small, he makes errors; when zero, he is (by chance) correct about 50 per cent of the time. (2) If the source is predominantly periodic, as with pure

270

tones or multiple tones, the curve itself is periodic, as shown in *Figure 1b* for a pure 800 c/s sine wave. These two forms of curve represent extreme tendencies; real-life sources of sound contain both (quasi-) periodic and stochastic components.

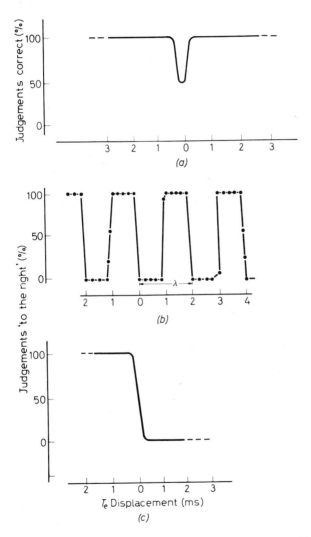

Figure 1. Fusion judgment curves for random and periodic sources: *(a)* lateralization for a white-noise source, speech source, or a similar quasi-random source; *(b)* lateralization for a periodic 800 c/s source; *(c)* replotting of curve *(a)* in terms of percentage of judgments to the right

271

We shall concentrate first on sources of type 1—for example, speech or noise.

The near-Gaussian dip in *Figure 1a* represents the subject's uncertainty, which has at least two origins: (1) since the fused image has a finite subjective size, its central point is uncertain, and (2) the subject is uncertain of his own midplane (intracranial) about which he is making acoustic judgments.

The arbitrariness between zero values of T_e on the magnetic tape and the central (intracranial) plane is removed by replotting *Figure 1a* in terms of percentage of right-hand judgments, as in *Figure 1c*, thereby skewing the curve. This method is adopted hereafter for all forms of 'judgment curve'. The fact that $T_{e\,=\,0}$ now is not to be taken as 'an applied central image'. Note that *Figure 1b* is plotted this way also.

So much, briefly, for the case of identical aural stimuli. If, however, the left- and right-hand signals, $S_L(t)$ and $S_R(t)$, are utterly different (statistically independent), say, two different voices, or different noises, then no fusion takes place (Cherry[3]). What we are here regarding as a 'fusion mechanism' is not triggered. This mechanism is then operated, or not operated, as determined by the relation between $S_L(t)$ and $S_R(t)$.

It has been our purpose to examine experimentally the nature of this critical relation and to find a model for the analytical processes whereby the brain determines it, as a control for the 'fusion mechanism'.

Obviously it is far more interesting to see what happens when $S_L(t)$ and $S_R(t)$ are neither identical nor totally unlike, but only *partially* alike. In what sense is 'alike' to be defined? What does the brain look for in its running analysis?

If we were to confine ourselves to the use of simple determinate signals, like sine waves, clicks, and so forth, we might oversimplify the problem. For in real life the aural stimuli, such as human speech, street noises, and so on, are stochastic. 'Stochastic' means 'probabilistic'; the brain cannot know in advance exactly what is coming to the ears next, in microscopic detail, instant by instant. Further than this, the two signals $S_L(t)$ and $S_R(t)$ are never *exactly* alike, in detail, in real life. They differ not only in timing T_e and average intensity A_L/A_R, owing to the spatial directions of sound sources, but in other ways, owing to head sound-shadows, reflections and reverberations, and random sound contributions from the wind on our faces.

Again, in real life, an aural image stays fused and does not hop about as the two aural signals fluctuate. It must be average, statistical, invariant properties that control this 'fusion mechanism', and several different storages may be used, for different averaging times.

Mathematically speaking, the measures that assess the average (statistical) degree of dependence, or independence, of stochastic sources are the *correlation functions*. There are different forms of such measures, but certain other real-life conditions, under which the brain must operate, narrow down the possibilities. Such functions have found a considerable place in models of other aural phenomena, notably in the work of Licklider[4, 5], especially in connection with pitch perception and spatial separation, but not, so far as we can find, in models of binaural fusion prior to our 1956 paper (Cherry and Sayers[2]). We shall be returning to such theoretical aspects in a later section.

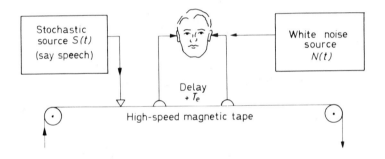

Figure 2. Control of the statistical dependence between the two-ear stimuli

EXPERIMENTAL CONTROL OF THE STATISTICAL INDEPENDENCE OF THE TWO-EAR SIGNALS

Before theorizing further, let us refer to some experiments. *Figure 2* shows the same binaural stimulus arrangement as resulted in the curves of *Figure 1*, except that now band-limited white noise is added to the signal at one ear only. The magnitude of this noise $N(t)$ in relation to that of $S(t)$ controls the statistical independence of the two-ear stimuli to any degree we wish.

Figure 3a shows a typical judgment curve (percentage of left judgments) as T_e is set to successive random values when $S(t)$ is male speech. This curve differs from that of *Figure 1* in two ways: (1) Dips appear, showing less certainty of left laterali-

273

zation of the image at certain values of time T_e. Fusion is now only *partial* over various regions of interaural delay T_e. (2) A total left-right dissymmetry shows *total fusion* when the pure speech signal leads in time but *partial fusion* when the noisy speech leads. This we have referred to as a type of *precedence effect* (Cherry and Sayers[2]).

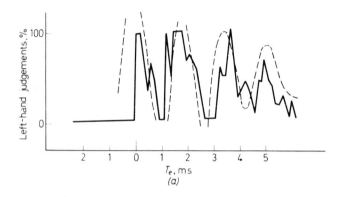

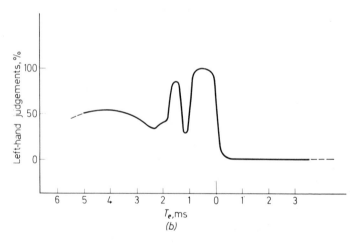

Figure 3. The binaural fusion 'precedence effect': *(a)* typical judgment curve using male speech masked by noise in left ear only [$S_R(t)$=pure speech; $S_L(t)$=noisy speech]; *(b)* typical judgment curve, using noise for left ear and noisy noise for right ear (theoretical curve dashed)

This new 'precedence effect' seems to us to have important implications, and it will be referred to again in the next section.

My colleague Dr. Sayers has made extensive measurements of such binaural 'judgment curves', each the result of one to two thousand separate right-left judgments with randomly set time delays. Partly from such curves the nature of the analysis performed by the brain (in the 'fusion mechanism') has been exposed. This analysis is required, in part, to determine the degree of statistical independence of the signals that arrive at each ear in real life, since this assessed degree of independence is a measure of the likelihood that these two signals come from a single external source of sound. If this likelihood is higher than some threshold (not necessarily constant), fusion is enforced; if lower, separate images are set up at each side of the head.

If the logic of such argument is accepted, we are led to the conclusion that the 'fusion mechanism' can be described as carrying out some type of correlation analysis, though there can by many forms of such measures. To determine the precise form used by the 'fusion mechanism', further experiments were performed using not only stochastic signals like speech and noise but determinate ones such as sinusoids. We shall return to these experiments later, in the section on Binaural Fusion of Speech. But first let us return for a moment to the curious 'precedence effect', already noted.

THE 'PRECEDENCE' ASSUMED BY A PURE SOUND SOURCE AT ONE EAR OVER A NOISY OR DISTORTED SOURCE AT THE OTHER EAR

There are other 'precedence effects' in hearing, but the one referred to earlier and illustrated in *Figure 3* has not been reported, so far as we are aware, prior to our 1956 and, in more detail, our 1957 papers (Cherry and Sayers[2], Sayers and Cherry[1]).

The effect was mentioned only in passing in these papers, and the writer would like now to take the opportunity of calling the attention of psychologists and neurophysiologists to it, since there may be important implications.

With reference to *Figure 3* the effect is this: if the 'pure' signal $S_R(t)$ (say, speech) is set earlier in time than the noisy signal $S_L(t)+N(t)$, then total fusion takes place, with zero uncertainty concerning its right-left localization, over a wide range of signal-to-noise ratios, provided the delay T_e is greater than some residual amount (about 100 μs in practice); but if the noisy signal is earlier in time, fusion is quite varied and uncertain. This total dissymmetry is illustrated by the judgment curve of *Figure 3a*.

In our earlier publications (1956, 1957) no explanation of this effect was offered. It was examined over a wide range of conditions, however, with not only white-noise masking in one ear but also various distortions (for example, severe amplitude clipping).

What distinguishes a 'pure' signal from a noisy or distorted one, thereby giving precedence to the former? Our first reaction was to assume that, in the case of speech, the listener's great experience of speech gives him prior information (and thus a set toward 'pure' speech). But experiments soon showed the effect to exist, equally strongly, with sources other than speech (such as combinations of sine waves of different frequency, and many others), of which we humans have far less prior knowledge. Finally, we can destroy all effects of prior knowledge by using, as the binaural source $S(t)$, another wide-band white-noise source, independent of the added noise source $N(t)$.*

Figure 3b shows a typical curve, which is quite asymmetrical about $T_e = 0$. Notice, in this case, that the 'noisy noise' $S(t)+N(t)$ has taken precedence over the 'pure noise' $S(t)$.

Clearly, then, the inference process by which the brain (fusion mechanism) discriminates between the two signals does not necessarily require prior probabilities (though with speech or other common signals such prior knowledge may conceivably enhance the effect). Recently H. B. Voelcker (of our laboratory) has given a complete theoretical explanation of the effect; unfortunately, at the time of writing, this has not appeared in print, so no reference can be given. Without anticipating its appearance, we might merely point out that the signal $S_R(t)$ is 'contained within' the noisy or distorted signal $S_L(t)+N(t)$, but that the converse is not true. Voelcker has examined the nature of the relation 'contained within' mathematically, concluding that this precedence effect requires only a simple correlation-analysis mechanism, which can be of the type that we have described as the basis of binaurel fusion (Sayers and Cherry[1], and which is outlined in the section on The Model of the Binaural Fusion Process in the present paper. In other words he observes that the processes adopted in our model are all that are strictly needed to explain the effect.

---

* Perhaps we could give a psychological definition to white noise, as being 'that signal of which the subject can have no prior microscopic knowledge (but only statistical knowledge), which prevents him from making predictions for a time ahead greater than the Nyquist Sampling Time.'

BINAURAL FUSION OF SPEECH

A great deal of auditory analysis is carried out with simple determinate signals, like clicks and pure sine waves. Speech, which is the class of signal of greatest importance in real life, has a very complex structure, but this does not preclude its use for such experiments as we are considering here. On the contrary, we regard as essential the use of speech and other stochastic signals, when attempting to build models of aural mechanisms—which, after all, have evolved around such natural sources.

The usual and most effective way of dealing with such complex 'stochastic' sources as speech is to use statistical (average) methods and measures. The method we have adopted, for studying binaural fusion, is of the type illustrated by *Figures 1* and *3*; our original paper (Sayers and Cherry[1]) contains many more.

It is from the parameters of such statistical data that the parameters of the binaural-fusion mechanism may be inferred. But such parameters must not be expected necessarily to remain invariant as the type of signal is changed (or the noise of the environment or other conditions). This variation we have found to be the case, in certain ways. But it is not the fusion *mechanism* that seems to change; rather it is extremely flexible in the way it handles a limitless variety of binaurel signals—speech, with its transients, its quasi-random breath noises, its quasi-periodic formants as well as all the myriad of other sound sources that surround us in daily life. Rather it is the *data* that the fusion mechanism recognizes as being 'in common' between the two ears that are found to vary according to circumstances; if one class of data is not available in the stimuli, the mechanism finds another.

This kind of flexibility is, of course, indicative also of correlation processes being used by the brain. Before examining for the exact processes, let us illustrate this flexibility in the case of speech.

THE DATA IN SPEECH CONTROLLING THE BINAURAL FUSION PROCESS

In this series of experiments, intoned vowels were used (to remove transients). They were recorded on tape, which was subsequently closed into a loop to provide a continuous source.

When such an intoned vowel is presented binaurally to a subject, the resulting 'judgment curve' is exactly the same as that of *Figure 1a* for a random or quasi-random source (or re-plotted so as to eliminate arbitrariness of zero delay T_e as in *Figure 1c*). This is indicative of the fact that the random breath sounds, closely similar in both ears, control the fusion process; the formant periodicities are not apparent here.

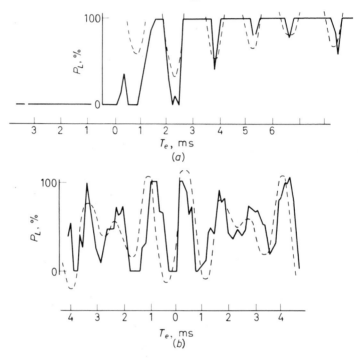

Figure 4. Judgment curves for *(a)* the intoned vowel *a*, noise-masked in the left ear only, with signal-to-noise ratio −9 db (theoretical curve dashed); *(b)* two sinusoids (600 and 800 c/s, equal level)

But we can easily destroy the similarity of the breath sounds arriving at each ear by adding to one side a white-noise generator, as in *Figure 2*. What does the mechanism fuse on now? *Figure 4a* shows one example, with an average signal-to-noise ratio of −9 db, the difference now is that dips have appeared at values of T_e that indicate that fusion is taking place on the periodic components (formants) of the intoned vowel source (which have known frequencies). Thus the fusion mechanism is operated by common breath sounds, under quiet conditions, but partly by the common formants when breath tones are rendered useless.

We shall not repeat details of the calculations here but merely state that the correspondence between the dips in the curve of *Figure 4a* and the formant frequencies of the intoned vowel has been examined by Sayers (Sayers and Cherry[1]); his calculated 'judgment curve' is shown dashed on *Figure 4a*. Note the coincidence of the peaks. Such a calculation, based on the mathematical model to be described in the next section, is possible with some accuracy because the formant frequencies and decrements can be measured accurately in this case of a steadily intoned vowel.

The asymmetry of the curve of *Figure 4a* representing our 'precedence effect' should also be noted.

THE MODEL OF THE BINAURAL FUSION PROCESS

The few curves we have illustrated here (*Figures 1, 3, 4*) are typical of a very large number resulting from measurement with many different types of sound source and different listening subjects (Sayers and Cherry[1]).

Let us look now at the steps involved in building up a theoretical model, by which such fusion 'judgment curves' may be calculated for all sorts of sound sources—pure tones, chords, intoned vowels, noise, running speech.

The first 'simple model' of binaural fusion

The simplest interaural correlation process is illustrated by *Figure 5*. Briefly, we argue that the L and R signals, $S_L(t)$ and $S_R(t)$, are cross-correlated and that the correlation function $R_{12}(t, \tau)$ is represented upon a 'conceptual surface', as shown by the dashed curve. Then the 'judgment mechanism' decides whether the function lies, on an average, more to the left or more to the right of the mid- ('intracranial') line. The measure used does not appear to be critical, since only a dichotomous judgment is needed; for our calculations, we have used the normalized areas lying, left and right, under the curve $R_{12}(t, \tau)$. One additional operation is necessary at this stage; before cross-correlation, the signals $S_L(t)$ and $S_R(t)$ need to be combined with their own mean values A_L and A_R (the method of combination used tentatively at this stage is simple linear addition). The reason for this is that these mean intensities can themselves also control the R-L judgment and so must appear upon the 'conceptual surface'. This oversimplified operation is clarified in our final model, as described in a later section.

Conventional cross-correlation is represented by the function

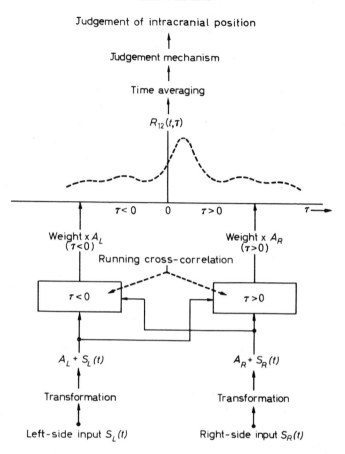

Figure 5. Model of the simple cross-correlation theory of binaural fusion. Note that the running cross-correlation unit has been divided in two parts only for clarity of representation

$$\Phi_{12}(\tau) = \lim_{T \to \infty} \frac{1}{2T} \int_{-T}^{+} f(t) \cdot g(t+\tau) \, dt \qquad \ldots . (1)$$

but clearly this is not usable, since the aural processes are proceeding in real time, and integration over the future is unrealistic. Again, the brain cannot integrate indefinitely over the past, because judgments are made moment by moment, as sources of sound move about. Clearly, the process can only be one of short-term *running correlation*, represented most generally by

$$R_{12}(t, \tau) = \int_{-\infty}^{+\infty} f(t-T)g(t-T-\tau)W(T)\,dt \qquad \cdots \cdot (2)$$

where $W(T)$ is a weighting function, corresponding to a short-term memory. The exact form of this function is not critical, we have found, but it is convenient to assume that it is exponential:

$$W(T) = \epsilon^{-KT}\Big|_{T > 0} \qquad \cdots \cdot (3)$$

Licklider (1951) used a similar running correlation function in his model of pitch perception. We have estimated the time constant K here as approximately 6 ms. This running cross-correlation is illustrated by *Figure 6*.

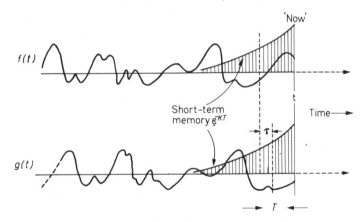

Figure 6. Short-term running correlation, as from equation 2

Theoretical results, based upon the 'simple model'

The key to the exact form of the correlation process was first obtained from fusion measurements based upon single and multiple pure tones (see, for example, *Figures 1b* and *4b*). Such curves appeared to be closely similar to the correlation functions of the signals themselves (for such steady-state signals no short-term memory ϵ^{-KT} was involved).

Notice one important thing about these curves, typical of all our measurements; they are 'sawn-off' sharply at the 100 per cent and 0 per cent levels, corresponding to the listener's absolute certainty of L or R. We shall cite only one example here: *Figure 4b* shows (dashed) the calculated curve based upon the 'simple model' when 600 and 800 c/s are applied together binaurally. This does not, of course, recognize the existence of these 100 per cent and 0 per cent limits, since the model contains no 'thresholds of absolute confidence'.

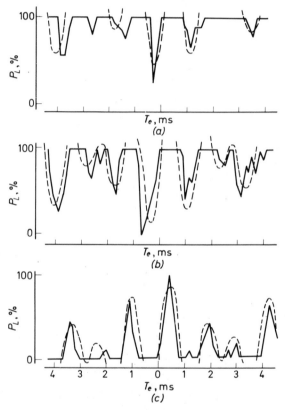

Figure 7. Judgment curves for two-sinusoid binaural signal, showing influence of interaural amplitude difference (600 c/s component, −2 db with respect to 800 c/s component in each channel). (a) $S_L(t)$ above $S_R(t)$ by 10 db; (b) by 6 db; and (c) by −6 db. Theoretical curves dashed

The complete model of binaural fusion

The 'simple model' of *Figure 5* was found to be inadequate for several reasons. Predominant is the fact that the L-R lateral movement of a fused sound image is controlled also by the relative *intensities* of the two ear signals A_L and A_R, as well as by their timing in relation to one another. We have made many measurements of fusion 'judgment curves', using, in particular, multiple sinusoids and intoned vowels, and have found the following a most surprising result.

Briefly, if the *form* of the signals $S_L(t)$ and $S_R(t)$ is held constant, but their relative mean intensities A_L/A_R are varied, the *form* of the judgment curve itself also remains invariant; however, this judgment curve is moved bodily, up and down, between the 'sawn-off' 100 per cent and 0 per cent limits. *Figure 7* shows one typical result, using two sinusoids, 600 and 800 c/s, as in *Figure 4b* also. It is as though the 100 and 0 per cent limits form a mask, behind which the judgment curve slides up and down, as controlled by the ratio A_L/A_R.

Such a result is very simply accounted for if the relative signal amplitudes A_L and A_R at the left and right ears are assessed, by some averaging process, and used to weight the L and R sides of the correlation function $R_{12}(t, \tau)$ (see *Figure 5*) accordingly, before the final judgment process decides whether this function lies predominantly L or R of the mid- (intracranial) line.

Such an assessment of A_L and A_R might well result from an additional *autocorrelation* process, operating on the signals at each ear separately. Again, if this is short term, as for the cross-correlation $R_{12}(t, \tau)$ in equation 2, running average values are assessed, $A_L(t)$ and $A_R(t)$, so that these are themselves functions of time. Such a method of assessment provides, inherently, a theoretical basis for other binaural phenomena, to be referred to later.

The complete model upon which calculations have been based is now shown in *Figure 8* and is described in detail in our 1957 paper. Briefly, the signals at each ear $S_L(t)$ and $S_R(t)$ are first subjected to separate running autocorrelation, and the two resulting functions are then cross-correlated.

This cross-correlation function $R_{12}(t, \tau)$ is subsequently represented on a 'conceptual surface' but, before the 'judgment mechanism' decides whether this function lies predominantly L or R of the midline, the two parts of the function (L and R) are weighted by a long-term time average of the two autocorrelation functions, $R_{11_L}(t, \tau)$ and $R_{11_R}(t, \tau)$. There are, therefore, three 'conceptual surfaces' involved, one at each ear and one in the centre.

Such a model is, of course, not neurophysiological, but *functional;* it has been used for calculating L-R lateral 'judgment curves', with many forms of binaural signal, either identical at each ear or noise-masked at one ear. We cannot cite many here (other than *Figures 3a, 4b*, and *7*) but would refer the interested reader to the original (Sayers and Cherry[1]).

SOME FURTHER AURAL EFFECTS PREDICTED BY THE MODEL

1. When the left or right ear only, but not both, is stimulated, no centrally fused subjective image is formed. In the model, only the L or R 'conceptual surface', and not the central one, is energized. Similarly, when the L and R signals are from statistically independent sources the same result obtains (Cherry[3]).

2. When a complex source of sound applied binaurally has no frequency components below approximately 1,200 to 1,500 c/s, it is an experimental fact that only the *envelopes* of the L and R signals fuse into a binaural image; their microstructures do not operate the fusion mechanism (Leakey, Sayers, and Cherry[6]). But when the signals are identical (or different) *pure* tones above about 1,500 c/s, with *no* varying envelopes, no fused image forms. This phenomenon assists our directional hearing of signals, all of whose components of wave length are shorter than twice the distance between the ears. Our fusion model does the same, since, under these conditions, the running autocorrelation processes at each ear automatically pass on only the envelopes (running averages) to the central cross-correlation process.

3. When listening binaurally to multiple sinusoids, the subject can, at will, listen to the whole fused chord, or to any one, or any group, of fused tones as distinct *Gestalten*. Again, by independent time-delay controls, each of these different fused images can, subjectively, have independent lateral movement. Our model also enables us to calculate the 'judgment curves' of lateral image positions for each of the images separately (Sayers and Cherry[1]), since the preliminary autocorrelation processes perform the necessary periodicity discriminations.

4. If, with binaural stimulation, one ear is, in addition, masked by noise, or if other perturbation is introduced, a kind of 'precedence effect' arises which we noted in an earlier section. Voelcker has shown that this can be explained only in terms of the short-term auto- and cross-correlation processes used in the model, but his work is not yet published.

The relations between our model and the proposals made by Licklider[5] concerning pitch perception might be noted here. We have adopted the same running-correlation function as Licklider

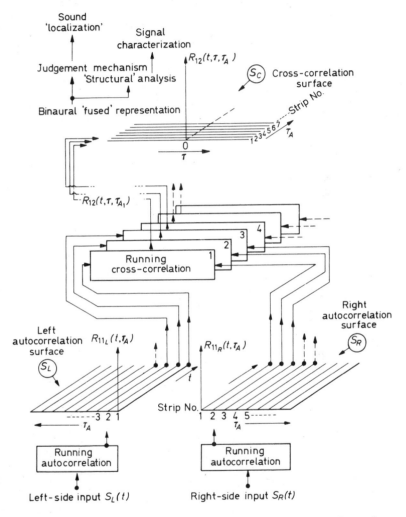

Figure 8. Model including the proposed extensions to the basic theory, from which calculations have been made

and, like him, we require both auto- and cross-correlation (although the order is reversed in our model). But our premises and experimental data are utterly different in kind. Licklider is

285

concerned with neurophysiological facts; we are not. Again, it is on pitch perception that he concentrates, rather than upon the various fusion phenomena. We have been concerned solely with building a functional model of fusion processes, that is, with a set of mathematical operations whereby we can *calculate numerically* the probability of binaural fusion—nothing else.

It cannot be overstressed that models such as ours are to be regarded as 'one of a class of models'; it is always possible to take the mathematical processes cited and recast them, or transform them, into alternative processes. Thus there is no *need* to restrict our processes to correlations; correlations can well be interpreted by filtering processes instead. And, once again, it is not neurophysiology that concerns us here, but representations of behaviour—entirely an 'outside view' of the human black box.

A great many perception phenomena concern extraction of *statistical invariants* from among the raw sense data; for *Gestalten* are recognized under many forms of transformation and perturbation—both determinate and statistical, for example, noise. Again, the perceptions operate in real time, moment by moment.

Short-term correlation thus seems to be a logical choice for the extraction of such invariants. We end then with a somewhat vague suggestion that perhaps the various hierarchical processes of perception may be described by hierarchical successions of such correlations. The question is whether short-term correlations may not be the bricks of which part of the house of perception is built.

SUMMARY

In this paper a mathematical model is developed which is descriptive of the binaural fusion process. Fusion of all types of signal can be handled: sinusoids, noise sounds, intoned vowels, running speech, and so forth.

It is entirely stimulus-response *behaviour* that is described by the model, and no reference whatever is made to the physiology or anatomy of hearing. With the model calculations can be made, in complete detail, of the probability with which a human listener will hear sounds as lying to the *right* or *left* side of his intracranial plane, when binaural stimuli are presented with some mutual time delay. The two signals (left-ear, right-ear) need not be identical; fusion still occurs when they are different, as in fact they always are in real life. Again, the mathematical model will assess the probability of fusion. Only if the two signals are totally unlike (statistically independent) are we assured of total failure of fusion.

The details of the model have already been published elsewhere. In the present paper attention is paid particularly to the arguments underlying the development of the model. Its relation to models of other hearing processes is clarified.

Finally, some typical examples are cited, comparing the experimental and the calculated results, and some general implications of the model for binaural perception are drawn.

REFERENCES

[1] Sayers, B. McA., and Cherry, E. C., (1957). 'Mechanism of binaural fusion in the hearing of speech.' *J. acoust. Soc. Am.*, *29*, 973-987

[2] Cherry, E. C., and Sayers, B. McA., (1956). 'Human "cross-correlator"—A technique for measuring certain parameters of speech perception.' *J. acoust. Soc. Am.*, *28*, 889-895

[3] Cherry, E. C., (1953). 'Some experiments upon the recognition of speech, with one and with two ears.' *J. acoust. Soc. Am.*, *25*, 975-979.

[4] Licklider, J. C. R., (1951). 'A duplex theory of pitch perception.' *Experientia*, *7*, 128-134

[5] Licklider, J. C. R., (1956). Auditory frequency analysis. In C. Cherry (Editor), *Proceedings of the Third London Symposium on Information Theory*. London; Butterworth

[6] Leakey, D. M., Sayers, B. McA. and Cherry, E. C., (1958). 'Binaural fusion of low- and high-frequency sounds.' *J. acoust. Soc. Am.*, *30*, 222-223

16

OBSERVATIONS ON MR BABBAGE'S ANALYTICAL ENGINE

(Editorial comment continued from page 1.)

A good mathematician, but not a particularly level-headed one, Lady Lovelace, later bankrupted herself on a progressive gambling system, but even so we give her pride of place in this book, for apart from using her footnotes to the first section, we also include as the epilogue her famous comments or 'objections' on computers which have philosophical interest even today.

OBSERVATIONS ON MR BABBAGE'S ANALYTICAL ENGINE

Lady A. A. Lovelace

The Analytical Engine has no pretensions whatever to *originate* anything. It can do whatever we *know how to order it* to perform. It can *follow* analysis; but it has no power of *anticipating* any analytical relations or truths. Its province is to assist us in making *available* what we are already acquainted with. This it is calculated to effect primarily and chiefly of course, through its executive faculties; but it is likely to exert an *indirect* and reciprocal influence on science itself in another manner. For, in so distributing and combining the truths and the formulae of analysis, that they may become most easily and rapidly amenable to the mechanical combinations of the engine, the relations and the nature of many subjects in that science are necessarily thrown into new lights, and more profoundly investigated. This is a decidedly indirect, and a somewhat speculative, consequence of such an invention. It is, however, pretty evident, on general principles, that in devising for mathematical truths a new form in which to record and throw themselves out for actual use, views are likely to be induced, which should again react on the more theoretical phase of the subject. There are in all extensions of human power, or additions to human knowledge, various *collateral* influences, besides the main and primary object attained.